Functional Calculus - Recent Advances and Development

Edited by Hammad Khalil

Published in London, United Kingdom

Functional Calculus - Recent Advances and Development
http://dx.doi.org/10.5772/intechopen.100721
Edited by Hammad Khalil

Contributors
M'hamed Elomari, Ali El Mfadel, Louay S. Yousuf, Xiaojing Shen, Yiwei Liao, Junfeng Wang, Yunmin Zhu, Carlos Granados, Guillermo J. Flores, Hammad Khalil, Tehseen Zahra, Azeem Shahzad, Zaffer Elahi

First published in London, United Kingdom, 2023 by IntechOpen
IntechOpen is the global imprint of INTECHOPEN LIMITED, registered in England and Wales, registration number: 11086078, 5 Princes Gate Court, London, SW7 2QJ, United Kingdom

British Library Cataloguing-in-Publication Data
A catalogue record for this book is available from the British Library

Additional hard and PDF copies can be obtained from orders@intechopen.com

Functional Calculus - Recent Advances and Development
Edited by Hammad Khalil
p. cm.
Print ISBN 978-1-80356-332-9
Online ISBN 978-1-80356-333-6
eBook (PDF) ISBN 978-1-80356-334-3

Meet the editor

Dr. Hammad Khalil is an assistant professor of mathematics in the Department of Mathematics at the University of Education, Lahore. He completed his Ph.D. in applied mathematics under the supervision of Professor Dr. Rahmat Ali Khan. He has published numerous articles in various international journals and serves on a number of editorial boards. His research areas include functional analysis, approximation theory, chaos theory and nonlinear dynamics.

Contents

Preface

Calculus is the elementary subject of applied analysis and its study includes a rich variety of functions and their behavior. This book brings together a range of different concepts from across the wide spectrum of the concept of calculus.

The book is in two sections. The first deals with advances in analysis and the second with the application of some results of functional calculus to applied problems.

The first section opens with an analysis of logarithmic potential transform. The singular values of this transform are discussed on the Poincaré disk. This potential can be used to illustrate some of the important features of field theory such as dimensional regularization and renormalization. Although most recent textbooks do not discuss this potential in detail, the calculations to demonstrate some of its unique features are quite simple. The bound state energy of this logarithmic potential is obtained through the uncertainty principle, phase space quantization and the Hellmann–Feynman theorem.

In the second chapter of the first section, the authors define and prove new Tauberian theorems under triple statistically Norlund-Cesaro summability. Some theorems, lemmas and corollaries can be defined and proved similarly by using the (1, 0, 0), (0, 1, 0) and (0, 0, 1) summability method. Although Tauberian theorems for single sequences of a single variable are well established, they remain in their infancy for triple sequences.

The final chapter of the first section is devoted to the study of the Calderon operator, which is the sum of the Hardy averaging operator and its adjoint and plays an important role in the theory of real interpolation. On the other hand, the Hilbert operator arises from the continuous version of Hilbert's inequality. Both operators appear in different contexts and have numerous applications within the harmonic analysis. In this chapter, the authors briefly review the Calderon and Hilbert operators, showing some of the most relevant results within the functional analysis and presenting recent results on these operators within Fourier analysis.

The second section of the book collects some results from applied analysis The first chapter deals with the study of heat transfer development of titanium oxide nanofluid of platelet-shaped nanoparticles over a vertical stretching cylinder. A set of nonlinear equations is obtained using suitable transformation on the governing equations which are then solved with numerical scheme BVP4C. The results obtained are interpreted graphically and numerically. The effects of Prandtl, Eckert and unsteadiness parameters on temperature distribution are depicted, and skin friction and Nusselt number are also computed. In the second chapter, a nonlinear response of the follower motion is simulated at different cam speeds, different coefficients of restitution and different internal distances of the follower guide from inside. The nonlinear response of the follower is employed to investigate the chaotic phenomenon in the cam follower system in the presence of follower offset. The numerical results are achieved using Solid Works software. The chaos phenomenon is detected using Poincaré maps with phase-plane portraits, the largest Lyapunov exponent parameter, and a bifurcation diagram. The largest Lyapunov exponent has its maximum value when the follower offsets to the right, and its minimum value when the follower offsets to the left. The chaotic phenomenon in cam follower systems when the follower offsets to the left is greater than the chaotic phenomenon when the follower offsets to the right. The final chapter investigates the decision fusion problem for large-scale sensor networks associated with the Internet of Things and artificial intelligence. The sensor networks discussed are those with unavoidable transmission channel interference and non-ideal channels that are prone to errors. A generalized algorithm is proposed that enables decision fusion rules to be designed for large-scale sensor networks and can at the same time search for the optimal sensor rules and the optimal fusion rule. Finally, numerical examples show the effectiveness of the new algorithms for large-scale sensor networks with non-ideal channels.

Hammad Khalil
Department of Mathematics,
University of Education,
Lahore, Pakistan

Section 1

Advances in Theory

Chapter 1

The Singular Values of the Logarithmic Potential Transform on Bound States Spaces of Landau Hamiltonians on the Poincaré Disk

M'hamed Elomari and Ali El Mfadel

Abstract

In the present manuscript, we prove that the singular numbers of the Cauchy transform $\mathcal{L}_\sigma[f](z) = -\frac{1}{\pi}\int_{\mathbb{D}} \frac{f(\xi)}{\xi - z} \log\left(\frac{1}{|z-\xi|}\right)\left(1 - \xi\overline{\xi}\right)^{\sigma-2} d\mu(\xi)$ (2) defined on the space $L^{2,\sigma}(\mathbb{D})$ of complex-valued measurable functions, which are $\left(1 - \xi\overline{\xi}\right)^{\sigma-2} d\mu(\xi)$-square integrable on $\mathbb{D}$ where $\sigma > 1$ is a fixed parameter, are asymptotically $\approx C\sqrt{k^{m-4\nu+1}}$, as $k \to \infty$ where C is a constant.

Keywords: the logarithmic potential transform, the singular values, Cauchy transform.

1. Introduction

Let $\mathbb{D}$ be the complex unit disk endowed with its Lebesgue measure μ and let ∂ID be its boundary. Denote by $L^2(\mathbb{D}, d\mu)$ the space of complex-valued measurable functions, which are $d\mu$ square integrable on $\mathbb{D}$. The logarithmic potential transform: $L^2(\mathbb{D}) \to L^2(\mathbb{D})$ is defined by

$$\mathcal{L}[f](z) = -\frac{1}{\pi}\int_{\mathbb{D}} \frac{f(\xi)}{\xi - z} \log\left(\frac{1}{|z-\xi|}\right) d\mu(\xi). \tag{1}$$

This operator is very important as the transformed Cauchy and it often appears in Analysis [1].

The dimensional analysis [1, 2] and scaling arguments form an integral part of theoretical physics to solve some important problems without doing much calculation.

The logarithmic potential in physics forms an interesting one as it provides some unusual predictions about the system. Moreover, this potential can be used suitably to illustrate some of the important features of field theory such as dimensional regularization and renormalization. In most of our textbooks, this potential is not discussed in detail; although the calculations are quite simple to demonstrate some of its unique features. We have obtained the bound state energy of this logarithmic potential through uncertainty principle, phase space quantization, and the Hellmann-Feynman theorem.

In Ref. [3] the authors have been dealing with the restriction of $\mathcal{L}$ to the space $L^2_a(\mathbb{D})$ of analytic μ-square integrable on $\mathbb{D}$. They precisely have considered the projection operator P_0: $L^2(\mathbb{D}) \to L^2_a(\mathbb{D})$ and they have proved that the singular values λ_k of $\mathcal{L}P_0$, (which turn out to be eigenvalues of the operator $\sqrt{(\mathcal{L}P_0)^*(\mathcal{L}P_0)}$ behave like k^{-1} as k goes to ∞. They also concluded that $\mathcal{L}P_0$ belongs to the Schatten class $S_{1,\infty}$.

Now, consider the following *weighted* logarithmic potential transform

$$\mathcal{L}_\sigma[f](z) = -\frac{1}{\pi}\int_{\mathbb{D}} \frac{f(\xi)}{\xi - z}\, log\left(\frac{1}{|z-\xi|}\right)(1-\xi\overline{\xi})^{\sigma-2} d\mu(\xi), \tag{2}$$

defined on the space $L^{2,\sigma}(\mathbb{D})$ of complex-valued measurable functions, which are $(1-\xi\overline{\xi})^{\sigma-2}d\mu(\xi)$-square integrable on $\mathbb{D}$ where $\sigma > 1$ is a fixed parameter. We observe that the subspace $L^{2,\sigma}_a(\mathbb{D})$ of analytic functions on $\mathbb{D}$ and belonging to $L^{2,\sigma}(\mathbb{D})$ coincides with the eigenspace

$$\mathcal{A}^\sigma_0(\mathbb{D}) := \{\psi \in L^{2,\sigma}(\mathbb{D}),\ \Delta_\sigma\psi = 0\}, \tag{3}$$

of the second order differential operator

$$\Delta_\sigma := -4(1-z\overline{z})\left((1-z\overline{z})\frac{\partial^2}{\partial z\partial\overline{z}} - \sigma\overline{z}\frac{\partial}{\partial\overline{z}}\right), \tag{4}$$

known as the σ-weight Maass Laplacian and its discrete eigenvalues are given by

$$\varepsilon_m := 4m(\sigma - 1 - m),\ m = 0,1,2,\dots,\lfloor(\sigma-1)/2\rfloor, \tag{5}$$

with their corresponding eigenspaces

$$\mathcal{A}^\sigma_m(\mathbb{D}) := \{\psi \in L^{2,\sigma}(\mathbb{D}) \text{ and } \Delta_\sigma\psi = \varepsilon^\sigma_m\psi\}, \tag{6}$$

are here called *generalized Bergman* spaces since ...

After noticing that, we here deal with analogous questions as in Ref. [3] in the context of the weighted Cauchy transform (2) and for its restriction to the space $\mathcal{A}^\sigma_m(\mathbb{D})$. That is, we are concerned with the operator $\mathcal{C}_\sigma P^\sigma_m$ where P^σ_m is the projection $L^{2,\sigma}(\mathbb{D}) \to \mathcal{A}^\sigma_m(\mathbb{D})$. The results achieved are as follows:

Firstly, we find that the singular values of $\mathcal{L}_\sigma P^\sigma_m$. For $k \neq m$, it can be expressed as

$$\lambda_k = \sqrt{J_1 + J_2 + J_3},$$

where

$$J_1 = \left(\frac{(1+k-m)_m}{m!(k-m+1)}\right)^2 \sum_{n=0}^{\infty} A_n \frac{\Gamma(2n+2k-2m+6-1)\Gamma(4\nu-2m-1)}{\Gamma(2n+2k-4m+4\nu+6)},$$

$$J_2 = \left(\frac{\alpha^{\nu,m}_k}{2\nu-m-1}\right)^2 \sum_{n=0}^{\infty} A_n \frac{\Gamma(4\nu-2m-1)\Gamma(2n+2)}{\Gamma(2n+4\nu-2m+1)},$$

and

$$J_3 = \frac{(1+k-m)_m \alpha_k^{\nu,m}}{m!(k-m+1)(2\nu-m-1)} \left(\sum_{n=0}^{\infty} A_n \frac{\Gamma(k-m+2)\Gamma(4\nu-2m-1)}{\Gamma(4\nu-k-3m)} \right).$$

For $k=m$ can be expressed as

$$\lambda_k^2 = \frac{\alpha_k^{\nu,m}(2(2\nu-m)-1)}{8(\pi(2\nu-m+1))} \sum_{n=0}^{\infty} \frac{B_n}{n+2\nu-m}, \tag{7}$$

where

$$B_n = \sum_{n=0}^{\infty} \frac{\Gamma(-m+1)\Gamma(2\nu-m)\Gamma(2(\nu-m)+1)}{n!\Gamma(2(\nu-m))\Gamma(2\nu-m+2)},$$

$$\alpha_k^{\nu,m} = \frac{\Gamma(2)\Gamma(2(m-\nu)+1)}{\Gamma(m+1)\Gamma(2+m-2\nu)}.$$

Secondly, we show that these singular values behave like

$$\lambda_k \sim C\sqrt{k^{m-4\nu+1}}, \text{ as } k \to \infty,$$

where C is a constant.

The paper is organized as follows: In Section 2, we review the definition of the weighted logarithmic potential transform, as well as some of its needed properties. Section 3 deals with some basic facts on the spectral theory of Mass Laplacians on the Poincaré disk. In Section 4, a precise description of the generalized Bergmann spaces is reviewed. Section 5 is devoted to the computation of the singular values of the weighted logarithmic potential transform. The asymptotic behavior of these singular values is established in Section 6.

2. The weighted logarithmic potential transform $\mathcal{L}_\nu$

2.1 The case $\nu = 1$

Let $\mathbb{D}$ the complex unit disk endowed with its Lebesgue measure μ and let ∂ID its boundary denote by $L^2(\mathbb{D})$ the space of complex-valued measurable functions on $\mathbb{D}$ with finite norm

$$\|f\| = \int_{\mathbb{D}} |f(\xi)|^2 d\mu(\xi). \tag{8}$$

The Logarithmic Potential operator $\mathcal{L} : L^2(\mathbb{D}) \to L^2(\mathbb{D})$ is defined by

$$\mathcal{L}[f](z) = \int_{\mathrm{ID}} f(\xi) \log\left(\frac{1}{|\xi - z|}\right) d\mu(\xi). \tag{9}$$

2.2 The case of $\nu \geq 1$

We fix a real parameter ν such that $2\nu > 1$ and we consider the following weighted logarithmic potential transform

$$\mathcal{L}_\nu[f](z) = \int_{\mathbb{D}} f(\xi) \log\left(\frac{1}{|\xi - z|}\right) (1 - \xi\overline{\xi})^{2\nu-2} d\mu(\xi), \tag{10}$$

defined on the space $L^{2,\nu}(\mathbb{D})$ complex-valued measurable functions are $(1 - \xi\overline{\xi})^{2\nu-2} d\mu(\xi)$-square integrable on $\mathbb{D}$. As a convolution of $L^{2,\nu}$-functions with the compactly supported measure $\frac{(1-\xi)^{2\nu-2}}{\xi} \mathbb{1}_{\mathrm{D}} d\mu(\xi)$ $\mathcal{L}_\nu : L^{2,\nu}(\mathbb{D}) \to L^{2,\nu}(\mathbb{D})$ is obviously bounded. Moreover, it is not hard to show that $\mathcal{L}_\nu$ is in fact compact [4]. This raises a question concerning the spectral picture of $\mathcal{L}_\nu$.

3. The Landau Hamiltonian H_ν on the Poincaré disk $\mathbb{D}$

Let $\mathbb{D} = \{z \in \mathbb{C}, z\overline{z} < 1\}$ be the complex unit disk with the Poincaré metric $ds^2 = 4(1 - z\overline{z})^{-2} dz d\overline{z}$. $\mathbb{D}$ is a complete Riemannian manifold with all sectional curvature equal -1. It has an ideal boundary ∂ID identified with the circle $\{\omega \in \mathbb{C}, \omega\overline{\omega} = 1\}$. One refers to points $\omega \in \partial$ID as points at infinity. The geodesic distance between two points z and w is given by

$$\cosh d(z, w) = 1 + \frac{2(z - w)(\overline{z} - \overline{w})}{(1 - z\overline{z})(1 - w\overline{w})}. \tag{11}$$

By Ref. [5] the Schrödinger operator on $\mathbb{D}$ with a constant magnetic field of strength proportional to $\nu > 0$ can be written as:

$$\mathcal{L}_\nu := -\left(1 - |z|^2\right)^2 \frac{\partial^2}{\partial z \partial \overline{z}} - \nu z\left(1 - |z|^2\right) \frac{\partial}{\partial z} + \nu \overline{z}\left(1 - |z|^2\right) \frac{\partial}{\partial \overline{z}} + \nu^2 |z|^2. \tag{12}$$

which is also called Maass Laplacian on the disk. A slight modification of $\mathcal{L}_\nu$ is given by the operator

$$H_\nu := 4\mathcal{L}_\nu - 4\nu^2 \tag{13}$$

acting in the Hilbert space

$$L^{2,0}(\mathbb{D}) := \left\{\varphi : \mathbb{D} \to \mathbb{C}, \int_{\mathbb{D}} |\varphi(z)|^2 \left(1 - |z|^2\right)^{-2} d\mu(z) < +\infty\right\}, \tag{14}$$

For our purpose, we shall consider the unitary equivalent realization $\tilde{H}_\nu$ of the operator H_ν in the Hilbert space

$$L^{2,\nu}(\mathbb{D}) := \left\{\varphi : \mathbb{D} \to \mathbb{C}, \int_{\mathbb{D}} |\varphi(z)|^2 \left(1 - |z|^2\right)^{2\nu-2} d\mu(z) < +\infty\right\}, \tag{15}$$

which is defined by

$$\tilde{H}_\nu := Q_\nu^{-1} H_\nu Q_\nu, \tag{16}$$

where $Q_\nu : L^{2,\nu}(\mathbb{D}) \to L^{2,0}(\mathbb{D})$ is the unitary transformation defined by the map $\varphi \mapsto Q_\nu[\varphi] := \left(1 - |z|^2\right)^{-\nu} \varphi$. Different aspects of the spectral analysis of the operator $\tilde{H}_\nu$ have been studied by many authors. For instance, note that $\tilde{H}_\nu$ is an elliptic densely defined operator on the Hilbert space $L^{2,\nu}(\mathbb{D})$ and admits a unique self-adjoint realization that we denote also by $\tilde{H}_\nu$. The spectrum of $\tilde{H}_\nu$ in $L^{2,\nu}(\mathbb{D})$ consists of two parts: (i) a continuous part $[1, +\infty[$, which corresponds to *scattering states*, (ii) a finite number of eigenvalues (*hyperbolic Landau levels*) of the form

$$\varepsilon_m^\nu := 4(\nu - m)(1 - \nu + m), \; m = 0,1,2,\cdots,\left[\nu - \frac{1}{2}\right] \tag{17}$$

with infinite degeneracy, provided that $2\nu > 1$. The eigenvalues in (17) correspond eigenfunctions, which are called *bound states* since the particle in such a state cannot leave the system without additional energy. A concrete description of these bound states spaces will be the goal of the next section.

4. The bound states spaces $\mathcal{A}^2_{\nu,m}(\mathbb{D})$

Here, we consider the eigenspace

$$\mathcal{A}^2_{\nu,m}(\mathbb{D}) := \left\{\Phi : \mathbb{D} \to \mathbb{C},\ \Phi \in L^{2,\nu}(\mathbb{D}) \text{ and } \tilde{H}_\nu \Phi = \varepsilon_m^\nu \Phi\right\}. \tag{18}$$

See Refs. [6, 7], for the following proposition.

Proposition 4.1. *Let* $2\nu > 1$ *and* $m = 0,1,2,\cdots,\left[\nu - \frac{1}{2}\right]$.*Then, we have.*
(i) *an orthogonal basis of* $\mathcal{A}^2_{\nu,m}(\mathbb{D})$ *is given by the functions*

$$\begin{aligned}\phi_k^{\nu,m}(z) &:= |z|^{|m-k|}\left(1 - |z|^2\right)^{-m} e^{-i(m-k)\arg z} \\ &\times {}_2F_1\left(-m + \frac{m-k+|m-k|}{2},\ 2\nu - m + \frac{|m-k| - m + k}{2},\ 1+|m-k|;\ |z|^2\right)\end{aligned} \tag{19}$$

$k = 0,1,2,\cdots$, *in terms of a terminating* ${}_2F_1$ *Gauss hypergeometric function.*
(ii) *the norm square of* $\phi_k^{\nu,m}$ *in* $L^{2,\nu}(\mathbb{D})$ *is given by*

$$\left\|\phi_k^{\nu,m}\right\|^2 = \frac{\pi(\Gamma(1 + |m-k|))^2}{(2(\nu - m) - 1)} \frac{\Gamma\left(m - \frac{|m-k|+m-k}{2} + 1\right)\Gamma\left(2\nu - m - \frac{|m-k|+m-k}{2}\right)}{\Gamma\left(m + \frac{|m-k|-m+k}{2} + 1\right)\Gamma\left(2\nu - m + \frac{|m-k|-m+k}{2}\right)}. \tag{20}$$

Corollary 4.1. *The functions* $\left\{\Phi_k^{\nu,m}\right\}$, $k = 0,1,2,\ldots$, *given by*

$$\Phi_k^{\nu,m}(z) := (-1)^k \left(\frac{2(\nu - m) - 1}{\pi}\right)^{\frac{1}{2}} \left(\frac{k!\Gamma(2(\nu - m) + m)}{m!\Gamma(2(\nu - m) + k)}\right)^{\frac{1}{2}} \tag{21}$$

$$\times \left(1 - |z|^2\right)^{-m} \bar{z}^{m-k} P_k^{(m-k,\ 2(\nu-m)-1)}(1 - 2z\bar{z}), \tag{22}$$

in terms of Jacobi polynomials constitute an orthonormal basis of $\mathcal{A}^{2,\nu}_m(\mathbb{D})$.

Proof. Write the connection between the ${}_2F_1$-sum and the Jacobi polynomial

$$P_k^{\alpha,\beta}(u) = \frac{(1+\alpha)_k}{k!} \cdot {}_2F_1\left(-k, 1+\alpha+\beta+k, 1+\alpha; \frac{1-u}{2}\right),$$

then the functions

$$\phi_k^{\nu,m}(z) = \frac{(-1)^{\min(m,\,k)}}{\left(1-|z|^2\right)^m} |z|^{|m-k|} e^{-i(m-k)\arg z} P_{\min(m,\,k)}^{(|m-k|,\, 2(\nu-m)-1)}(1-2z\bar{z}), \tag{23}$$

constitute an orthonormal basis of $\mathcal{A}_{\nu,m}^2$. The norm square of $\phi_k^{\nu,m}$ in $L^{2,\nu}(\mathbb{D})$ is given by

$$\left\|\phi_k^{\nu,m}\right\|^2 = \frac{\pi}{(2(\nu-m)-1)} \frac{(m\vee k)!\Gamma(2(\nu-m)+m\wedge k)}{(m\wedge k)!\Gamma(2(\nu-m)+m\vee k)}. \tag{24}$$

Here, $m\wedge k := \min(m, k)$ and $m\vee k := \max(m, k)$. Thus, the set of functions

$$\Phi_k^{\nu,m} := \frac{\phi_k^{\nu,m}}{\left\|\phi_k^{\nu,m}\right\|}, k = 0,1,2,\ldots \tag{25}$$

is an orthonormal basis of $\mathcal{A}_{\nu,m}^2(\mathbb{D})$ and can be rewritten as.

$$\Phi_k^{\nu,m}(z) = (-1)^k \left(\frac{2(\nu-m)-1}{\pi}\right)^{\frac{1}{2}} \left(\frac{k!\Gamma(2(\nu-m)+m)}{m!\Gamma(2(\nu-m)+k)}\right)^{\frac{1}{2}} \tag{26}$$

$$\times\left(1-|z|^2\right)^{-m} \bar{z}^{m-k} P_k^{(m-k,\, 2(\nu-m)-1)}(1-2z\bar{z}) \tag{27}$$

by making appeal to the identity $(S, p.63)$:

$$\frac{\Gamma(m+1)}{\Gamma(m-s+1)} P_m^{(-s,\,\alpha)}(u) = \frac{\Gamma(m+\alpha+1)}{\Gamma(m-s+\alpha+1)} \left(\frac{u-1}{2}\right)^s P_{m-s}^{(s,\,\alpha)}(u), 1 \le s \le m \tag{28}$$

for $s = m-k$, $t = 1-2|z|^2$, and $\alpha = 2(\nu-m)-1.\ldots\square$.

Corollary 4.2. *The L^2–eigenspace $\mathcal{A}_{\nu,0}^2(\mathbb{D})$, corresponding to $m = 0$ in (3.1) and associated with the bottom energy $\varepsilon_0^\nu = 0$ in (2.6), reduces further to the weighted Bergman space consisting of holomorphic functions ϕ: $\mathbb{D} \to \mathbb{C}$ such that*

$$\int_{\mathbb{D}} |\phi(z)|^2 \left(1-|z|^2\right)^{2\nu-2} d\mu(z) < +\infty. \tag{29}$$

5. Computation of the singular values λ_k

Elements of this basis are given in terms of Jacobi polynomials as

$$\phi_k^{\nu,m}(z) = \frac{(-1)^{\min(m,\,k)}}{\left(1-|z|^2\right)^m}|z|^{|m-k|}e^{-i(m-k)\arg z}P_{\min(m,\,k)}^{(|m-k|,\,2(\nu-m)-1)}(1-2z\bar{z}). \tag{30}$$

The norm square of $\phi_k^{\nu,m}$ in $L^{2,\nu}(\mathbb{D})$ is given by

$$\rho_k^{\nu,m} = \frac{\pi}{(2(\nu-m)-1)}\frac{(m\vee k)!\Gamma(2(\nu-m)+m\wedge k)}{(m\wedge k)!\Gamma(2(\nu-m)+m\vee k)}. \tag{31}$$

Here, $m\wedge k := \min(m,k)$ and $m\vee k := \max(m,k)$. Let us introduce the notation. The set of functions

$$\gamma_k^{\nu,m} := \frac{(-1)^{m\wedge k}}{\sqrt{\rho_k^{\nu,m}}}, k = 0,1,2, \ldots \tag{32}$$

So that we consider the elements

$$\Phi_k^{\nu,m}(z) := \gamma_k^{\nu,m}\frac{1}{(1-z\bar{z})^m}|z|^{|m-k|}e^{-i(m-k)\arg z}P_{\min(m,\,k)}^{(|m-k|,\,2(\nu-m)-1)}(1-2z\bar{z}). \tag{33}$$

5.1 The action $\mathcal{L}_\nu$

Lemma 5.1. *We set* $z=\rho e^{it}$, *and* $I = -\int_0^{2\pi} e^{i(k-m)\theta}\log\left(\left|z-re^{i\theta}\right|\right)\frac{d\theta}{2\pi}$, *we have*

$$\begin{cases} I = -\log(\rho\wedge r) & k=m, \\ I = \dfrac{e^{i(k-m)t}}{2|m-k|}\left(\left(\dfrac{r}{\rho}\right)^{m-k}\wedge\left(\dfrac{r}{\rho}\right)^{m-k}\right) & k\neq m, \end{cases} \tag{34}$$

Proof. By ref. [3], it remains to prove that this lemma for $k<m$.
We have

$$\int_0^{2\pi} e^{i(k-m)\theta}\log\left(\left|\rho e^{it}-re^{i\theta}\right|\right)d\theta = -\int_0^{2\pi} e^{i(m-k)(-\theta)}\log\left(\left|re^{i(-t)}-\rho e^{i(-\theta)}\right|\right)d(-\theta) \tag{35}$$

The function $\theta \to e^{i(m-k)(-\theta)}\log\left(\left|re^{i(-t)}-\rho e^{i(-\theta)}\right|\right)$ is a periodic mapping with the period equal 2π, then

$$\int_0^{2\pi} e^{i(k-m)\theta} \log\left(\left|\rho e^{it} - re^{i\theta}\right|\right)d\theta$$
$$= -\int_0^{2\pi} e^{i(m-k)(-\theta)} \log\left(\left|re^{i(-t)} - \rho e^{i(-\theta)}\right|\right)d(-\theta)$$
$$= \frac{e^{i(k-m)t}}{2(m-k)} \times \left(\left(\frac{r}{\rho}\right)^{m-k} \wedge \left(\frac{r}{\rho}\right)^{m-k}\right).$$

□

Lemma 5.2. *For all $\lambda \in \partial D$. $\mathcal{L}_\nu$ commutes with the rotations R_λ, where*

$$(R_\lambda f)(z) = f(\lambda z).$$

Proof. We observe that

$$R_\lambda \phi_k^{\nu,m}(z) = \lambda^{k-m} \phi_k^{\nu,m}(z), \ \forall k \neq m.$$

□

Corollary 5.1. $\left\{\mathcal{L}_\nu\left(\phi_k^{\nu,m}\right)\right\}_{k=0}^{\infty}$ *are orthonormal in $L^{2,\nu}(\mathbb{D})$.*

Proof. As R_λ is an isometry of $L^{2,\nu}(\mathbb{D})$,

$$\left(\mathcal{L}_\nu\left(\phi_k^{\nu,m}\right), \mathcal{L}_\nu\left(\phi_j^{\nu,m}\right)\right)$$
$$= \left(R_\lambda \mathcal{L}_\nu\left(\phi_k^{\nu,m}\right), R_\lambda \mathcal{L}_\nu\left(\phi_j^{\nu,m}\right)\right)$$
$$= \left(\mathcal{L}_\nu R_\lambda\left(\phi_k^{\nu,m}\right), \mathcal{L}_\nu R_\lambda\left(\phi_j^{\nu,m}\right)\right)$$
$$= \overline{\lambda^{j-k}}\left(\mathcal{L}_\nu\left(\phi_k^{\nu,m}\right), \mathcal{L}_\nu\left(\phi_j^{\nu,m}\right)\right), \text{ if } m > k,$$

or

$$= \lambda^{k-j}\left(\mathcal{L}_\nu\left(\phi_k^{\nu,m}\right), \mathcal{L}_\nu\left(\phi_j^{\nu,m}\right)\right), \text{ if } m < k$$

For all $\lambda \in \partial D$, since $\lambda \neq 0$, we have

$$\left(\mathcal{L}_\nu\left(\phi_k^{\nu,m}\right), \mathcal{L}_\nu\left(\phi_j^{\nu,m}\right)\right) = 0 \text{ if } j \neq k.$$

□

Lemma 5.3. *If we denote ${}^1\phi_k^{\nu,m}(z)$, if $k > m$ and ${}^2\phi_k^{\nu,m}(z)$, if $k < m$, we have*

$$\mathcal{L}_\nu\left({}^1\phi_k^{\nu,m}\right)(z) = \frac{\Gamma(k+1)\Gamma(2\nu - m)}{\Gamma(m+1)\Gamma(2\nu - k)} \mathcal{L}_\nu\left({}^2\phi_k^{\nu,m}\right)(z).$$

Proof. Just use

$$\frac{\Gamma(m+1)}{\Gamma(m-s+1)} P_m^{(-s,\,\alpha)}(u) = \frac{\Gamma(m+\alpha+1)}{\Gamma(m-s+\alpha+1)} \left(\frac{u-1}{2}\right)^s P_{m-s}^{(s,\,\alpha)}(u), 1 \leq s \leq m. \qquad (36)$$

□

Proposition 5.1. *The action of the operator $\mathcal{L}$ on a basis element $\phi_k^{\nu,m}$ is of the form:*
If $k = m$, We put $z = \rho e^{i\theta}$ then

$$\mathcal{L}_\nu\left(\phi_k^{\nu,m}\right)(z) = \frac{\alpha_k^{\nu,m}}{2(2\nu - m + 1)}\sqrt{\frac{2(\nu - m) - 1}{\pi}}\left(1 - \rho^2\right)^{2\nu - m - 1} {}_3F_2\left(\begin{matrix} -m + 1, 2\nu - m, 2\nu - m + 1 \\ 2(\nu - m), 2\nu - m + 2 \end{matrix} \middle| 1 - \rho^2\right). \tag{37}$$

If $k \neq m$ then

$$\mathcal{L}_\nu\left(\phi_k^{\nu,m}\right)(z) = \frac{\pi \gamma_k^{\nu,m} e^{i(k-m)t}}{2(k - m)}(I_3 + I_4),$$

where

$$I_3 = \frac{(1 + k - m)_m}{m!(k - m + 1)}\rho^{k-m+2}\left(1 - \rho^2\right)^{2\nu - m - 1} {}_2F_1\left(\begin{matrix} -m + 1, 2(\nu - m) + k \\ 2 + k - m \end{matrix} \middle| \rho^2\right),$$

and

$$I_4 = \frac{\alpha_k^{\nu,m}}{2\nu - m - 1}\left(1 - \rho^2\right)^{2\nu - m - 1} {}_2F_1\left(\begin{matrix} -m + 1, 2\nu - m - 1 \\ 2(\nu - m), \end{matrix} \middle| \rho^2\right).$$

Proof. For $k = m$, we have

$$\begin{aligned}
\mathcal{L}_\nu\left(\phi_k^{\nu,m}\right)(z) &= \frac{(-1)^m}{\pi}\sqrt{\frac{2(\nu - m) - 1}{\pi}}\int_{\mathbb{D}}\left(1 - |\xi|^2\right)^{2\nu - m - 2} P_m^{(0,\,2(\nu-m)-1)}\left(1 - 2|\xi|^2\right)\log\left(|z - \xi|\right)d\mu(\xi) \\
&= (-1)^m\sqrt{\frac{2(\nu - m) - 1}{\pi}}\int_0^1\left(1 - r^2\right)^{2\nu - m - 2} P_m^{(0,\,2(\nu-m)-1)}\left(1 - 2r^2\right)\log\left(\rho \wedge r\right)dr^2 \\
&= \frac{(-1)^m}{2}\sqrt{\frac{2(\nu - m) - 1}{\pi}}\int_0^1 (1 - t)^{2\nu - m - 2} P_m^{(0,\,2(\nu-m)-1)}(1 - 2t)\log\left(\rho^2 \vee t\right)dt \\
&= \frac{(-1)^m}{2}\sqrt{\frac{2(\nu - m) - 1}{\pi}}[I_1 + I_2].
\end{aligned}$$

where

$$I_1 = \int_0^{\rho^2} (1 - t)^{2\nu - m - 2} P_m^{(0,\,2(\nu-m)-1)}(1 - 2t)\log\left(\rho^2 \vee t\right)dt,$$

and

$$I_2 = \int_{\rho^2}^1 (1 - t)^{2\nu - m - 2} P_m^{(0,\,2(\nu-m)-1)}(1 - 2t)\log(t)dt.$$

Calculus of I_1.

$$I_1 = \log\left(\rho^2\right)\int_{\rho^2}^1 (1 - t)^{2\nu - m - 2} P_m^{(0,\,2(\nu-m)-1)}(1 - 2t)dt.$$

We use the formula

$$P_k^{(\alpha,\,\beta)}(u) = \frac{(1+\alpha)_k}{k!}\,{}_2F_1\left(\begin{matrix}-k,\,1+\alpha+\beta+k\\1+\alpha\end{matrix}\,\bigg|\frac{1-u}{2}\right).$$

We have

$$I_1 = \log\left(\rho^2\right)\int_0^{\rho^2}(1-t)^{2\nu-m-2}\,{}_2F_1\left(\begin{matrix}-m,\,2\nu-m\\1\end{matrix}\,\bigg|t\right)dt.$$

By Ref. [8], we have

$$\int x^{c-1}(1-x)^{b-c-1}\,{}_2F_1\left(\begin{matrix}a,\,b\\c\end{matrix}\,\bigg|x\right)dx = \frac{1}{c}x^c(1-x)^{b-c}\,{}_2F_1\left(\begin{matrix}a+1,\,b\\c+1\end{matrix}\,\bigg|x\right),$$

implies that

$$I_1 = \log\left(\rho^2\right)\rho^2\left(1-\rho^2\right)^{2\nu-m-1}\,{}_2F_1\left(\begin{matrix}-m+1,\,2\nu-m\\2\end{matrix}\,\bigg|\rho^2\right).$$

Calculus of I_2.

$$I_2 = \int_{\rho^2}^1(1-t)^{2\nu-m-2}P_m^{(0,\,2(\nu-m)-1)}(1-2t)\log(t)dt.$$

Use the previous formula in Ref. [8] and the integration by part gives

$$\begin{aligned}I_2 &= \left[t1-t^{2\nu-m-1}\,{}_2F_1\left(\begin{matrix}-m+1,2\nu-m\\2\end{matrix}\,\bigg|t\right)\log t\right]_{\rho^2}^1 - \int_{\rho^2}^1 1-t^{2\nu-m}\,{}_2F_1\left(\begin{matrix}-m+1,2\nu-m\\2\end{matrix}\,\bigg|t\right)dt\\ &= -\rho^2\log\rho^2 1-\rho^{2\,2\nu-m-1}\,{}_2F_1\left(\begin{matrix}-m+1,\,2\nu-m\\2\end{matrix}\,\bigg|\rho^2\right) - \int_{\rho^2}^1 1-t^{2\nu-m}\,{}_2F_1\left(\begin{matrix}-m+1,\,2\nu-m\\2\end{matrix}\,\bigg|t\right)dt.\end{aligned}$$

Calculus of

$$\int_{\rho^2}^1(1-t)^{2\nu-m}\,{}_2F_1\left(\begin{matrix}-m+1,\,2\nu-m\\2\end{matrix}\,\bigg|t\right)dt.$$

Use the following formula, which has place in [9]

$$\begin{aligned}{}_2F_1\left(\begin{matrix}a,\,b\\c\end{matrix}\,\bigg|t\right) &= \frac{\Gamma(c)\Gamma(c-a-b)}{\Gamma(c-a)\Gamma(c-b)}\,{}_2F_1\left(\begin{matrix}a,\,b\\a+b-c+1\end{matrix}\,\bigg|1-t\right)\\ &+ \frac{\Gamma(c)\Gamma(a+b-c)}{\Gamma(a)\Gamma(b)}(1-t)^{c-a-b}\,{}_2F_1\left(\begin{matrix}a,\,b\\a+b-c+1\end{matrix}\,\bigg|1-t\right).\end{aligned}$$

We put $a = 1-m$, $b = 2\nu-m$, $c = 2$ and use the formula Boher-Mollerup, for $z \in \mathrm{IR}_+^*$,

$$\Gamma(z) = \frac{e^{-\gamma z}}{z}\prod_{n=1}^{\infty}\left(1+\frac{z}{n}\right)^{-1}e^{-\frac{z}{n}},$$

which implies $\frac{1}{\Gamma(1-m)} = 0$, then

$$ {}_2F_1\left(\begin{matrix} -m+1,\, 2\nu-m \\ 2 \end{matrix} \middle| t\right) = \frac{2\Gamma(2(m-\nu)+1)}{m!\Gamma(2+m-2\nu)_2} F_1\left(\begin{matrix} -m+1,\, 2\nu-m \\ 2(\nu-m) \end{matrix} \middle| 1-t\right), $$

implies that

$$ \int_{\rho^2}^1 (1-t)^{2\nu-m} {}_2F_1\left(\begin{matrix} -m+1,\, 2\nu-m \\ 2 \end{matrix} \middle| t\right) dt = \frac{2\Gamma(2(m-\nu)+1)}{m!\Gamma(2+m-2\nu)} \int_{\rho^2}^1 (1-t)^{2\nu-m} {}_2F_1\left(\begin{matrix} -m+1,\, 2\nu-m \\ 2(\nu-m) \end{matrix} \middle| 1-t\right) dt. $$

By the change $1-t=s$, we get

$$ \int_{\rho^2}^1 (1-t)^{2\nu-m} {}_2F_1\left(\begin{matrix} -m+1,\, 2\nu-m \\ 2 \end{matrix} \middle| t\right) dt = \frac{2\Gamma(2(m-\nu)+1)}{m!\Gamma(2+m-2\nu)} \int_0^{1-\rho^2} t^{2\nu-m} {}_2F_1\left(\begin{matrix} -m+1,\, 2\nu-m \\ 2(\nu-m) \end{matrix} \middle| t\right) dt. $$

In [8], p. 44,

$$ \int x^{\alpha-1} {}_2F_1\left(\begin{matrix} a,\, b \\ c \end{matrix} \middle| -t\right) dx = \frac{x^\alpha}{\alpha}\, {}_3F_2\left(\begin{matrix} a,\, b,\, \alpha \\ c,\, \alpha+1 \end{matrix} \middle| -t\right) + \frac{\Gamma(\alpha)\Gamma(a-\alpha)\Gamma(b-\alpha)\Gamma(c)}{\Gamma(a)\Gamma(b)\Gamma(c-\alpha)} $$

Since $a=1-m$, $b=2\nu-m$, $c=2(\nu-m)$, and $\alpha=2\nu-m+1$ we have

$$ \frac{\Gamma(\alpha)\Gamma(a-\alpha)\Gamma(b-\alpha)\Gamma(c)}{\Gamma(a)\Gamma(b)\Gamma(c-\alpha)} = 0 $$

and by the change $t=-s$

$$ \begin{aligned} &\int_0^{1-\rho^2} t^{2\nu-m+1} {}_2F_1\left(\begin{matrix} -m+1,\, 2\nu-m \\ 2(\nu-m) \end{matrix} \middle| |t|\right) dt \\ &= (-1)^m \int_0^{\rho^2} t^{2\nu-m} {}_2F_1\left(\begin{matrix} -m+1,\, 2\nu-m \\ 2(\nu-m) \end{matrix} \middle| |-t|\right) dt \end{aligned} $$

$$ = (-1)^m \frac{(\rho^2-1)^{2\nu-m+1}}{2\nu-m+1}\, {}_3F_2\left(\begin{matrix} -m+1,\, 2\nu-m,\, 2\nu-m+1 \\ 2(\nu-m),\, 2\nu-m+2 \end{matrix} \middle| |1-\rho^2|\right). $$

we set $\alpha_k^{\nu,m} = \frac{2\Gamma(2(m-\nu)+1)}{m!\Gamma(2+m-2\nu)}$. We get

$$ I_2 = \begin{matrix} -\rho^2 \log\rho^2 1-\rho^{2\,2\nu-m-1} {}_2F_1\left(\begin{matrix} -m+1, 2\nu-m \\ 2 \end{matrix} \middle| \rho^2\right) \\ + -1^m \alpha_k^{\nu,m} \frac{1-\rho^{2\,2\nu-m-1}}{2\nu-m+1}\, {}_3F_2\left(\begin{matrix} -m+1, 2\nu-m, 2\nu-m+1 \\ 2\nu-m, 2\nu-m+2 \end{matrix} \middle| 1-\rho^2\right). \end{matrix} $$

Finally

$$ \mathcal{L}_\nu(\phi_k^{\nu,m})(z) = \frac{\alpha_k^{\nu,m}}{2(2\nu-m+1)} \sqrt{\frac{2(\nu-m)-1}{\pi}} (1-\rho^2)^{2\nu-m-1} {}_3F_2\left(\begin{matrix} -m+1,\, 2\nu-m,\, 2\nu-m+1 \\ 2(\nu-m),\, 2\nu-m+2 \end{matrix} \middle| 1-\rho^2\right). $$

Now if $k > m$, set $z = \rho e^{it}$.

$$\mathcal{L}_\nu\left(\phi_k^{\nu,m}\right)(z) = \gamma_k^{\nu,m}\int_{\mathbb{D}}\left(1-|\xi|^2\right)^{2\nu-m-2}\xi^{k-m}\log\left(\frac{1}{|z-\xi|}\right)P_m^{(k-m,\ 2(\nu-m)-1)}\left(1-2|\xi|^2\right)d\mu(\xi)$$

$$= \gamma_k^{\nu,m}\int_0^1\left(1-r^2\right)^{2\nu-m-2}r^{k-m+1}P_m^{(k-m,\ 2(\nu-m)-1)}\left(1-2r^2\right)\int_0^{2\pi}e^{i(k-m)\theta}\log\left(\frac{1}{|z-r^{i\theta}|}\right)d\theta dr$$

$$= \frac{\pi\gamma_k^{\nu,m}e^{i(k-m)t}}{2(k-m)}\int_0^1\left(1-r^2\right)^{2\nu-m-2}r^{k-m}P_m^{(k-m,\ 2(\nu-m)-1)}\left(1-2r^2\right)\left(\left(\frac{r}{\rho}\right)^{k-m}\wedge\left(\frac{\rho}{r}\right)^{k-m}\right)dr^2$$

$$= \frac{\pi\gamma_k^{\nu,m}e^{i(k-m)t}}{2(k-m)}\left(\int_0^\rho\left(1-r^2\right)^{2\nu-m-2}r^{k-m}P_m^{(k-m,\ 2(\nu-m)-1)}\left(1-2r^2\right)\left(\left(\frac{r}{\rho}\right)^{k-m}\wedge\left(\frac{\rho}{r}\right)^{k-m}\right)dr^2\right.$$

$$\left.+\int_\rho^1\left(1-r^2\right)^{2\nu-m-2}r^{k-m}P_m^{(k-m,\ 2(\nu-m)-1)}\left(1-2r^2\right)\left(\left(\frac{r}{\rho}\right)^{k-m}\wedge\left(\frac{\rho}{r}\right)^{k-m}\right)dr^2.\right.$$

We set

$$I_3 = \int_0^\rho\left(1-r^2\right)^{2\nu-m-2}r^{k-m}P_m^{(k-m,\ 2(\nu-m)-1)}\left(1-2r^2\right)\left(\left(\frac{r}{\rho}\right)^{k-m}\wedge\left(\frac{\rho}{r}\right)^{k-m}\right)dr^2.$$

and

$$I_4 = \int_\rho^1\left(1-r^2\right)^{2\nu-m-2}r^{k-m}P_m^{(k-m,\ 2(\nu-m)-1)}\left(1-2r^2\right)\left(\left(\frac{r}{\rho}\right)^{k-m}\wedge\left(\frac{\rho}{r}\right)^{k-m}\right)dr^2.$$

Calculus of I_3.

$$I_3 = \frac{\rho^{m-k}(1+k-m)_m}{m!}\int_0^{\rho^2}t^{k-m}(1-t)^{2\nu-m-2}{}_2F_1\left(\begin{matrix}-m,\ 2(\nu-m)+k\\ 1+k-m\end{matrix}\ \middle|\ t\right)dt.$$

By the formula

$$\int x^{c-1}(1-x)^{b-c-1}{}_2F_1\left(\begin{matrix}a,\ b\\ c\end{matrix}\ \middle|\ x\right)dx = \frac{1}{c}x^c(1-x)^{b-c}{}_2F_1\left(\begin{matrix}a+1,\ b\\ c+1\end{matrix}\ \middle|\ x\right),$$

we have

$$I_3 = \frac{(1+k-m)_m}{m!(k-m+1)}\rho^{k-m+2}\left(1-\rho^2\right)^{2\nu-m-1}{}_2F_1\left(\begin{matrix}-m+1,\ 2(\nu-m)+k\\ 2+k-m\end{matrix}\ \middle|\ \rho^2\right)$$

Calculus of I_4.

$$I_4 = \int_\rho^1\left(1-r^2\right)^{2\nu-m-2}r^{k-m}P_m^{(k-m,\ 2(\nu-m)-1)}\left(1-2r^2\right)\left(\left(\frac{r}{\rho}\right)^{k-m}\wedge\left(\frac{\rho}{r}\right)^{k-m}\right)dr^2$$

$$= \frac{\rho^{k-m}(1+k-m)_m}{2m!}\int_{\rho^2}^1(1-t)^{2\nu-m-2}{}_2F_1\left(\begin{matrix}-m,\ 2(\nu-m)+k\\ 1+k-m\end{matrix}\ \middle|\ t\right)dt.$$

As the previous

$$\int_{\rho^2}^{1}(1-t)^{2\nu-m-2}{}_2F_1\left(\begin{matrix}-m,\,2(\nu-m)+k\\1+k-m\end{matrix}\,\middle|\,t\right)dt=\alpha_k^{\nu,m}\int_{\rho^2}^{1}(1-t)^{2\nu-m-2}{}_2F_1\left(\begin{matrix}-m+1,\,2\nu-m\\2(\nu-m)\end{matrix}\,\middle|\,1-t\right)dt$$

$$=(-1)^m\alpha_k^{\nu,m}\int_{\rho^2-1}^{0}t^{2\nu-m-2}{}_2F_1\left(\begin{matrix}-m+1,\,2\nu-m\\2(\nu-m)\end{matrix}\,\middle|\,t\right)dt$$

$$=\frac{\alpha_k^{\nu,m}}{2\nu-m-1}\left(1-\rho^2\right)^{2\nu-m-1}{}_3F_2\left(\begin{matrix}-m+1,\,2\nu-m,\,2\nu-m-1\\2(\nu-m),\,2\nu-m\end{matrix}\,\middle|\,1-\rho^2\right)$$

also

$${}_3F_2\left(\begin{matrix}-m+1,\,2\nu-m,\,2\nu-m-1\\2(\nu-m),\,2\nu-m\end{matrix}\,\middle|\,1-\rho^2\right)={}_2F_1\left(\begin{matrix}-m+1,\,2\nu-m-1\\2(\nu-m)\end{matrix}\,\middle|\,1-\rho^2\right)$$

Now if $k<m$. We have

$$\phi_k^{\nu,m}(z)=(-1)^k\sqrt{\frac{2(\nu-m)-1}{\pi}\frac{k!\Gamma(2(\nu-m)+m)}{m!\Gamma(2(\nu-m)+k)}}\left(1-|z|^2\right)^{-m}\bar{z}^{m-k}P_k^{(m-k,\,2(\nu-m)-1)}\left(1-2|z|^2\right).$$

By the formula

$$\frac{\Gamma(m+1)}{\Gamma(m-s+1)}P_m^{(-s,\,\alpha)}(u)=\frac{\Gamma(m+\alpha+1)}{\Gamma(m-s+\alpha+1)}\left(\frac{u-1}{2}\right)^s P_{m-s}^{(s,\,\alpha)}(u),1\le s\le m, \tag{38}$$

and put $s=m-k$ and $\alpha=2(\nu-m)-1$, we have

$$P_k^{(m-k,\,2(\nu-m)-1)}\left(1-2|z|^2\right)=\frac{m!\Gamma(k+\alpha+1)}{k!\Gamma(m+\alpha+1)}P_m^{(k-m,\,2(\nu-m)-1)}\left(1-2|z|^2\right),$$

substituting in the expression of $\phi_k^{\nu,m}(z)$, we get

$$\phi_k^{\nu,m}(z)=(-1)^m\sqrt{\frac{2(\nu-m)-1}{\pi}\frac{m!\Gamma(2(\nu-m)+k)}{k!\Gamma(2(\nu-m)+m)}}\left(1-|z|^2\right)^{-m}z^{k-m}P_m^{(k-m,\,2(\nu-m)-1)}\left(1-2|z|^2\right),$$

it is the same formula for $k>m$, which proves the same formula of $\mathcal{L}_\nu\left(\phi_k^{\nu,m}\right)(z)$ if $k>m$.

Remark 5.1. *By the previous formula in [9], we have*

$${}_2F_1\left(\begin{matrix}-m+1,\,2(\nu-m)+k\\2(\nu-m)\end{matrix}\,\middle|\,\rho^2\right)=\frac{k!\Gamma(2+k-m)}{\Gamma(1-2(\nu-m))_2}F_1\left(\begin{matrix}-m+1,\,2(\nu-m)+k\\2(\nu-m)\end{matrix}\,\middle|\,1-\rho^2\right).$$

5.2 The spectrum of $\mathcal{L}_\nu$

Proposition 5.2. *If $k\neq m$, then*

$$\lambda_k=\sqrt{J_1+J_2+J_3}.$$

where

$$J_1 = \left(\frac{(1+k-m)_m}{m!(k-m+1)}\right)^2 \sum_{n=0}^{\infty} A_n \frac{\Gamma(2n+2k-2m+6-1)\Gamma(4\nu-2m-1)}{\Gamma(2n+2k-4m+4\nu+6)},$$

$$J_2 = \left(\frac{\alpha_k^{\nu,m}}{2\nu-m-1}\right)^2 \sum_{n=0}^{\infty} A_n \frac{\Gamma(4\nu-2m-1)\Gamma(2n+2)}{\Gamma(2n+4\nu-2m+1)}.$$

and

$$J_3 = \frac{(1+k-m)_m \alpha_k^{\nu,m}}{m!(k-m+1)(2\nu-m-1)} \left(\sum_{n=0}^{\infty} A_n \frac{\Gamma(k-m+2)\Gamma(4\nu-2m-1)}{\Gamma(4\nu-k-3m)}\right),$$

If $k = m$ then

$$\lambda_k^2 = \frac{\alpha_k^{\nu,m}(2(2\nu-m)-1)}{8(\pi(2\nu-m+1))} \sum_{n=0}^{\infty} \frac{B_n}{n+2\nu-m}. \tag{39}$$

where

$$B_n = \sum_{n=0}^{\infty} \frac{\Gamma(-m+1)\Gamma(2\nu-m)\Gamma(2(\nu-m)+1)}{n!\Gamma(2(\nu-m)\Gamma(2\nu-m+2)}.$$

Proof. If $k \neq m$. We have

$$\left(\mathcal{L}_\nu\left(\phi_k^{\nu,m}\right)\right)(z) = \frac{\pi\gamma_k^{\nu,m}(I_3+I_4)}{2(k-m)} e^{i(k-m)t}.$$

We set $H = \left(L^2(\mathbb{D}), \left(1-|\xi|^2\right)^{2\nu-2} d\mu(\xi)\right)$, $I_3 = I_3(\rho)$, and $I_4 = I_4(\rho)$ we have

$$\begin{aligned}\lambda_k^2 &= \left\langle \mathcal{L}_\nu\left(\phi_k^{\nu,m}\right), \mathcal{L}_\nu\left(\phi_k^{\nu,m}\right)\right\rangle_H \\ &= \frac{\pi^2\gamma_k^{\nu,m}}{(k-m)} \int_0^1 (I_3(\rho)+I_4(\rho))^2 \rho d\rho.\end{aligned}$$

Calculus of $\int_0^1 (I_3(\rho))^2 \rho d\rho$.

$$I_3(\rho) = \frac{(1+k-m)_m}{m!(k-m+1)} \rho^{k-m+2}\left(1-\rho^2\right)^{2\nu-m-1} {}_2F_1\left(\begin{matrix}-m+1, 2(\nu-m)+k \\ 2+k-m\end{matrix} \middle| \rho^2\right).$$

Since

$${}_2F_1\left(\begin{matrix}-m+1, 2(\nu-m)+k \\ 2+k-m\end{matrix} \middle| \rho^2\right) = \sum_{n=0}^{\infty} \frac{(-m+1)_n(2(\nu-m)+k)_n}{(2+k-m)_n} \frac{\rho^{2n}}{n!},$$

then

$$(I_3(\rho))^2 = \left(\frac{(1+k-m)_m}{m!(k-m+1)}\right)^2 \sum_{n=0}^{\infty} A_n \rho^{2n}\left(1-\rho^2\right)^{4\nu-2m-2}.$$

where

$$A_n = \frac{1}{n!}\sum_{i=0}^{n} \frac{(-m+1)_i(-m+1)_{n-i}(2(\nu-m)+k)_i(2(\nu-m)+k)_{n-i}}{(2(\nu-m))_i(2(\nu-m))_{n-i}}.$$

Thus

$$J_1 = \int_0^1 (I_3(\rho))^2 \rho d\rho = \left(\frac{(1+k-m)_m}{m!(k-m+1)}\right)^2 \sum_{n=0}^{\infty} A_n \int_0^1 \rho^{2n+2k-2m+6-1}\left(1-\rho^2\right)^{4\nu-2m-1-1} d\rho.$$

Use the fact that

$$\int_0^1 t^{\alpha-1}(1-t)^{\beta-1}dt = \frac{\Gamma(\alpha)\Gamma(\beta)}{\Gamma(\alpha+\beta)},$$

implies

$$\int_0^1 (I_3(\rho))^2 \rho d\rho = \left(\frac{(1+k-m)_m}{m!(k-m+1)}\right)^2 \sum_{n=0}^{\infty} A_n \frac{\Gamma(2n+2k-2m+6-1)\Gamma(4\nu-2m-1)}{\Gamma(2n+2k-4m+4\nu+6)}. \tag{40}$$

Calculus of $\int_0^1 (I_4(\rho))^2 \rho d\rho$.
In the same,

$$J_2 = \int_0^1 (I_4(\rho))^2 \rho d\rho = \left(\frac{\alpha_k^{\nu,m}}{2\nu-m-1}\right)^2 \sum_{n=0}^{\infty} A_n \frac{\Gamma(4\nu-2m-1)\Gamma(2n+2)}{\Gamma(2n+4\nu-2m+1)}. \tag{41}$$

Calculus of $2\int_0^1 (I_3(\rho))(I_4(\rho))\rho d\rho$.

$$J_3 = 2\int_0^1 (I_3(\rho))(I_4(\rho))\rho d\rho = \frac{(1+k-m)_m \alpha_k^{\nu,m}}{m!(k-m+1)(2\nu-m-1)} \sum_{n=0}^{\infty} A_n \frac{\Gamma(k-m+2)\Gamma(4\nu-2m-1)}{\Gamma(4\nu-k-3m)}. \tag{42}$$

If $k = m$.
Since

$$\left({}_3F_2\left(\begin{matrix} -m+1,\ 2\nu-m,\ 2(\nu-m)+1 \\ 2(\nu-m),\ 2\nu-m+2 \end{matrix} \middle| 1-\rho^2\right)\right)^2 = \sum_{n=0}^{\infty} \frac{\Gamma(-m+1)\Gamma(2\nu-m)\Gamma(2(\nu-m)+1)}{n!\Gamma(2(\nu-m)\Gamma(2\nu-m+2)} \left(1-\rho^2\right)^n. \tag{43}$$

$$\lambda_k^2 = \frac{\alpha_k^{\nu,m}(2(2\nu-m)-1)}{8(\pi(2\nu-m+1))} \sum_{n=0}^{\infty} B_n \int_0^1 \left(1-\rho^2\right)^{n+2\nu-m-1} d\rho = \frac{\alpha_k^{\nu,m}(2(2\nu-m)-1)}{8(\pi(2\nu-m+1))} \sum_{n=0}^{\infty} \frac{B_n}{n+2\nu-m}, \tag{44}$$

where

$$B_n = \sum_{n=0}^{\infty} \frac{\Gamma(-m+1)\Gamma(2\nu-m)\Gamma(2(\nu-m)+1)}{n!\Gamma(2(\nu-m)\Gamma(2\nu-m+2)}.$$

6. Asymptotic behavior of singular values λ_k as $k \to \infty$

Proposition 6.1.

$$\lambda_k \sim C\sqrt{k^{m-4\nu+1}}, \ as \ k \to \infty,$$

where C is a constant.
Proof. If $k > m$, then

$$\lambda_k = \sqrt{J_1 + J_2 + J_3},$$

where

$$J_1 = \left(\frac{(1+k-m)_m}{m!(k-m+1)}\right)^2 \sum_{n=0}^{\infty} A_n \frac{\Gamma(2n+2k-2m+6-1)\Gamma(4\nu-2m-1)}{\Gamma(2n+2k-4m+4\nu+6)},$$

$$J_2 = \left(\frac{\alpha_k^{\nu,m}}{2\nu-m-1}\right)^2 \sum_{n=0}^{\infty} A_n \frac{\Gamma(4\nu-2m-1)\Gamma(2n+2)}{\Gamma(2n+4\nu-2m+1)}$$

and

$$J_3 = \frac{(1+k-m)_m \alpha_k^{\nu,m}}{m!(k-m+1)(2\nu-m-1)} \left(\sum_{n=0}^{\infty} A_n \frac{\Gamma(k-m+2)\Gamma(4\nu-2m-1)}{\Gamma(4\nu-k-3m)}\right).$$

The limit of λ_k as $k \to \infty$.
We use the formula

$$\frac{\Gamma(k+a)}{k=b} \sim k^{a-b}$$

we have

$$J_1 \sim \left(\frac{k^{-1-m}}{m!}\right)^2 \sum_{n=0}^{\infty} A_n \Gamma(4\nu-2m-1)(2k)^{2m-4\nu-1} \sim k^{-4\nu-1} 2^{2m-4\nu-1} \Gamma(4\nu-2m-1) \sum_{n=0}^{\infty} \frac{A_n}{m!}. \tag{45}$$

$$J_2 = \mathcal{O}_{k \sim \infty}(1) \tag{46}$$

In the same

$$J_3 \sim k^{m-4\nu+1} \frac{\alpha_k^{\nu,m} \Gamma(4\nu-2m-1)}{m!(2\nu-m-1)} \sum_{n=0}^{\infty} A_n. \tag{47}$$

Therefore

$$\lambda_k \sim C\sqrt{k^{m-4\nu+1}},$$

where C is a constant.

Author details

M'hamed Elomari* and Ali El Mfadel
Department of Mathematics, Sultan Moulay Slimane University, Beni Mellal, Morocco

*Address all correspondence to: amar.elo@gmail.com

References

[1] Langhaar HL. Dimensional Analysis and Theory of Models. New York: Wiley; 1951

[2] Barenblatt GI. Dimensional Analysis. New York: Gordon Breach Science Publishers; 1987

[3] Arazy J, Khavinson D. Spectral estimates of Cauchy's transform in $L^2(\Omega)$. Integral Equation and Operational Theory. 1992;**15**:901-9019

[4] Anderson JM, Khavinson D, Lomonosov V. Spectral properties of some integral operators arising in potential theory. The Quarterly Journal of Mathematics. 1992;**43**(4):387-407

[5] Ferapontov EV, Veselov AP. Integrable Schrodinger operators with magnetic fields: Factorization method on curved surfaces. Journal of Mathematical Physics. 2001;**42**:590-607

[6] Mourad EH. Classical and Quantum Orthogonal Polynomials in One Variable, Encyclopedia of Mathematics and Its Applications. Cambridge: Cambridge University Press; 2005

[7] Magnus W, Oberhettinger F, Soni RP. Formulas and Theorems for the Special Functions of Mathematical Physics. Berlin Heidelberg New York: Springer-Verlag; 1966

[8] Prudnikov AP, Brychkov YA, Marichev OI. Integrals and Series Volume 3 More Spacial Functions. New York: Gordon and Breach; 1990

[9] Karlsson PW, Srivastava HM, Manocha HL. A treatise on generating functions. Bulletin (New Series) of the American Mathematical Society. 1988;**19**(1):346-348

Chapter 2

Some Tauberian Theorems under Triple Statistically Nörlund-Cesáro Summability Method

Carlos Granados

Abstract

In this paper, we extend the notion presented by Braha (2020) in a higher dimension, we introduce the notion of $N_{p,q}^{n,m,g}C_{n,m,g}^{(1,1,1)}$-statistically convergence and show necessity and sufficiency conditions under which the existence of the limit $st\text{-}\lim_{n,m,g\to\infty} x_{n,m,g} = L$ follows from that $st\text{-}\lim_{n,m,g\to\infty} N_{p,q}^{n,m,g}C_{n,m,g}^{(1,1,1)} = L$. These conditions are one-sided or two-sided if $(x_{n,m,g})$ is a sequence of real or complex numbers, respectively.

Keywords: Nörlund-Cesro summability method, one-sided and two-sided Tauberian conditions, triple statistical convergence

1. Introduction

The concept of statistical convergence was introduced by Fast [1] and Steinhaus [2]. Besides, in this connection, Fridy [3] showed some relation to a Tauberian condition for the statistical convergence of (x_k). Subsequently, many researchers have worked in this area in several settings. For more recent works in this direction, one may refer to [4, 5]. Existing works in this field based on statistical convergence appears to have been restricted to real or complex sequences; however, Parida et al. [6] extended the idea for a locally convex Hausdorff topological linear space. Tauber [7] introduced the first Tauberian theorems for single sequences, that an Abel summable sequence is convergent with some suitable conditions. Later, a huge number of authors such as Landau [8], Hardy and Littlewood [9], and Schmidt [10] obtained some classical Tauberian theorems for Cessáro and Abel summability methods of single sequences. Recently, Braha [11] introduced some notions on statistical convergence by using the Nörlund-Cesáro summability method in a single sequence and proved some Tauberian theorems. In the last year, Canak and Totur [12], and Jena et al. [13] presented and studied several Tauberian theorems for single sequences. On the other hand, Knopp [14] obtained some classical type Tauberian theorems for Abel and $(C, 1, 1)$ summability methods of double sequences and proved that Abel and $(C, 1, 1)$ summability methods hold for the set of bounded sequences. Further, Moricz [15] proved some Tauberian theorems for Cesáro summable double sequences and deduced Tauberian theorems of Landau [16] and Hardy [17] type. Canak and Totur [18] have proved a Tauberian theorem for Cesáro summability of single integrals and also the alternative proofs of some classical type Tauberian theorems for the Cesáro summability of single integrals and later introduced by Parida et al. [6] for double integrals. Otherwise, the notion of $(C,1,1,1)$ summability of a triple sequence was originally introduced by Canak and Totur in 2016 [19]. Later, Canak et al. [20] studied some $(C,1,1,1)$ means of a statistical convergent triple sequence and gave some classical Tauberian theorems for a triple sequence that P-convergence follows from statistically $(C,1,1,1)$ summability under the two-sided boundedness conditions and slowly oscillating conditions in certain senses. Then, in 2020 Totur and Canak [21] proved Tauberian conditions under which convergence of triple integrals follows from $(C,1,1,1)$ summability. For more studies associated to the main topic of this paper, we refer the reader to [22–24].

Let $\left(p_{n,m,g}\right)$ and $\left(q_{n,m,g}\right)$ be any two non-negative real sequences with

$$R_{n,m,g}=\sum_{i=0}^{n}\sum_{j=0}^{m}\sum_{k=0}^{g}p_{i,j,k}q_{n-i,m-j,g-k}\neq 0\ \ ((n,m,g)\in\mathbb{N}\times\mathbb{N}\times\mathbb{N})$$

and $(C,1,1,1)$-Cesáro summability method. Let $(x_{n,m,g})$ be a sequence of real of complex numbers and set

$$N_{p,q}^{n,m,g}C_{n,m,g}^{(1,1,1)}=\frac{1}{R_{n,m,g}}\sum_{i=0}^{n}\sum_{j=0}^{m}\sum_{k=0}^{g}p_{i,j,k}q_{n-i,m-j,g-k}\frac{1}{i+1}\frac{1}{j+1}\frac{1}{k+1}\sum_{u=0}^{i}\sum_{v=0}^{j}\sum_{y=0}^{k}x_{u,v,y}$$

for $(n,m,g)\in\mathbb{N}\times\mathbb{N}\times\mathbb{N}$.

In this paper, we show necessary and sufficient conditions under which the existence of the limit $\lim_{n,m,g\to\infty}x_{n,m,g}=L$ follows from that of $\lim_{n,m,g\to\infty}N_{p,q}^{n,m,g}C_{n,m,g}^{(1,1,1)}=L$. These conditions are one-sided or two-sided if $(x_{n,m,g})$ is a sequence of real or complex numbers, respectively.

Given two non-negative sequences $\left(p_{n,m,g}\right)$ and $\left(q_{n,m,g}\right)$, the convolution $(p\star q)$ is defined by

$$R_{n,m,g}=(p\star q)_{n,m,g}=\sum_{i=0}^{n}\sum_{j=0}^{m}\sum_{k=0}^{g}p_{i,j,k}q_{n-i,m-j,g-k}=\sum_{i=0}^{n}\sum_{j=0}^{m}\sum_{k=0}^{g}p_{n-i,m-j,g-k}q_{i,j,k}$$

with $(C,1,1,1)$ we will denote the triple Cesáro summability method. Now, let $(x_{n,m,g})$ be a sequence, when $(p\star q)_{n,m,g}\neq 0$ for all $(n,m,g)\in\mathbb{N}\times\mathbb{N}\times\mathbb{N}$ the generalized Nörlund-Cesáro transform of the sequence $(x_{n,m,g})$ is the sequence $N_{p,q}^{n,m,g}C_{n,m,g}^{(1,1,1)}$ obtained by putting

$$N_{p,q}^{n,m,g}C_{n,m,g}^{(1,1,1)}=\frac{1}{(p\star q)_{n,m,g}}\sum_{i=0}^{n}\sum_{j=0}^{m}\sum_{k=0}^{g}p_{i,j,k}q_{n-i,m-j,g-k}\frac{1}{i+1}\frac{1}{j+1}\frac{1}{k+1}\sum_{u=0}^{i}\sum_{v=0}^{j}\sum_{y=0}^{k}x_{u,v,y}. \tag{1}$$

We say that the sequence $(x_{n,m,g})$ is generalized Nörlund-Cesáro summable to L determined by the sequences $\left(p_{n,m,g}\right)$ and $\left(q_{n,m,g}\right)$ (or simply summable $N_{p,q}^{n,m,g}C_{n,m,g}^{(1,1,1)}$) to L if

$$\lim_{n,m,g\to\infty} N_{p,q}^{n,m,g} C_{n,m,g}^{(1,1,1)} = L. \tag{2}$$

Throughout this paper, we will assume that the sequences $\left(p_{n,m,g}\right)$ and $\left(q_{n,m,g}\right)$ are satisfying the following conditions

$$q_{n,m,g} \geq 1, \quad \sum_{i=0}^{n}\sum_{j=0}^{m}\sum_{k=0}^{g} p_{i,j} \sim nmg, \quad (n,m,g) \in \mathbb{N}\times\mathbb{N}\times\mathbb{N}, \tag{3}$$

$$q_{\lambda_{n-i,m-j,g-k}} \leq 2q_{n-i,m-j,g-k}, \quad i = 1,2,\dots,n; j = 1,2,\dots,m; \lambda > 1k = 1,2,\dots; \lambda > 1, \tag{4}$$

$$q_{n-i,m-j,g-k} \leq 2q_{\lambda_{n-i,m-j,g-k}} \quad i = 1,2,\dots,\lambda_n; j = 1,2,\dots,\lambda_m; k = 1,2,\dots,; 0 < \lambda < 1, \tag{5}$$

where $\lambda_n = [\lambda n]$, $\lambda_m = [\lambda m]$ and $\lambda_g = [\lambda g]$. On the other hand, $a_{n,m,g} \sim b_{n,m,g}$ means that there are constants C, C_1 such that $a_{n,m,g} \leq Cb_{n,m,g} \leq C_1 a_{n,m,g}$. If

$$\lim_{n,m,g\to\infty} x_{n,m,g} = L \tag{6}$$

implies (2), then the method $N_{p,q}^{n,m,g} C_{n,m,g}^{(1,1,1)}$ is said to be regular. Nevertheless, the converse is not always true as can be seen in the following example:

Let us consider that $p_{n,m,g} = q_{n,m,g} = 1$ for all $(n,m,g) \in \mathbb{N}\times\mathbb{N}\times\mathbb{N}$. Besides, we define the following sequence $x - \left(x_{i,j,k}\right) = (-1)^{i+j+k}$, then we get

$$\frac{1}{(n+1)(m+1)(g+1)}\left|\sum_{i=0}^{n}\sum_{j=0}^{m}\sum_{k=0}^{g}\frac{1}{(i+1)(j+1)(k+1)}\sum_{u=0}^{i}\sum_{v=0}^{j}\sum_{y=0}^{k}(-1)^{u+v+y}\right|$$

$$\leq \frac{1}{(n+1)(m+1)(g+1)}\sum_{i=0}^{n}\sum_{j=0}^{m}\sum_{k=0}^{g}\frac{1}{(i+1)(j+1)(k+1)}\sum_{u=0}^{i}\sum_{v=0}^{j}\sum_{y=0}^{k}1 \to 1 \text{ as } n,m,g \to \infty.$$

and as we know, $x = \left(x_{i,j,k}\right)$ is not convergent. Notice that (6) can imply (2) under a certain condition, which is called a Tauberian conditions. Any theorem which states that convergence of a sequence follows from its $N_{p,q}^{n,m,g} C_{n,m,g}^{(1,1,1)}$ summability and some Tauberian conditions are said to be a Tauberian theorems for the $N_{p,q}^{n,m,g} C_{n,m,g}^{(1,1,1)}$ summability method.

Next, we will find some conditions under which the converse implication holds, for defined convergence. Exactly, we will prove under which conditions statistical convergence of sequences $\left(x_{n,m,g}\right)$, follows from statistically Nörlund-Cesáro summability method.

A sequence $\left(x_{n,m,g}\right)$ is weighted $N_{p,q}^{n,m,g} C_{n,m,g}^{(1,1,1)}$-statistically convergent to L if for every $\varepsilon > 0$,

$$\lim_{n,m,g\to\infty}\frac{1}{(p\star q)_{n,m,g}}\left|\left\{i,j,k \leq (p\star q)_{n,m,g} : \frac{1}{(p\star q)_{n,m,g}}\sum_{i=0}^{n}\sum_{j=0}^{m}\sum_{k=0}^{g} p_{i,j,k} q_{n-i,m-j,g-k}\right.\right.$$

$$\left.\left.\frac{1}{i+1}\frac{1}{j+1}\frac{1}{(k+1)}\sum_{u=0}^{i}\sum_{v=0}^{j}\sum_{y=0}^{k} x_{u,v,y} - L\right| \geq \varepsilon\right\}\right| = 0.$$

And we say that the sequence $\left(x_{n,m,g}\right)$ is statistically summable to L by the weighted summability method $N_{p,q}^{n,m,g}C_{n,m,g}^{(1,1,1)}$ if $st-\lim_{n,m,g} N_{p,q}^{n,m,g}C_{n,m,g}^{(1,1,1)} = L$. We will denote by $N_{p,q}^{n,m,g}C_{n,m,g}^{(1,1,1)}(st)$ the set of all sequences which are statistically summable, by the weighted summability method $N_{p,q}^{n,m,g}C_{n,m,g}^{(1,1,1)}$.

Theorem 1.1 Let $x = \left(x_{n,m,g}\right)$ be a sequence $N_{p,q}^{n,m,g}C_{n,m,g}^{(1,1,1)}$ summable to L, then the sequence $x = \left(x_{n,m,g}\right)$ is $N_{p,q}^{n,m,g}C_{n,m,g}^{(1,1,1)}$-statistically convergent to L, but not conversely.

Proof: The first part of the proof is obvious. To prove the second part, we will show the following example:

Let us define

$$x_{i,j,k} = \begin{cases} \sqrt{xyz}, & \text{for } i = n^2 \ j = m^2 \text{ and } k = g^2 \\ 0, & \text{otherwise} \end{cases}$$

and $p_{n,m,g} = 1 = q_{n,m,g}$. Under this conditions we obtain,

$$\frac{1}{(n+1)(m+1)(g+1)}|\{i,j,k \le n+1, m+1, g+1 :$$
$$|\frac{1}{(n+1)(m+1)(g+1)}\sum_{i=0}^{n}\sum_{j=0}^{m}\sum_{k=0}^{g}\frac{1}{P_{i,j,k}}\sum_{u=0}^{i}\sum_{v=0}^{j}\sum_{y=0}^{k}p_{u,v,y}x_{u,v,y} - 0| \ge \varepsilon\}|$$
$$\le \frac{\sqrt{(n+1)(m+1)(g+1)}}{(n+1)(m+1)(g+1)} \to 0.$$

On the other hand, for $i = n^2, j = m^2$ and $k = g2$, we have

$$\frac{1}{(n+1)(m+1)(g+1)}\sum_{i=0}^{n}\sum_{j=0}^{m}\sum_{k=0}^{g}\frac{1}{(i+1)(j+1)(k+1)}\sum_{u=0}^{i}\sum_{v=0}^{j}\sum_{y=0}^{k}x_{u,v,y} \to \infty,$$
$$as\ n,m,g \to \infty.$$

From last relation follows that $x = \left(x_{n,m,g}\right)$ is not $N_{p,q}^{n,m,g}C_{n,m,g}^{(1,1,1)}$ summable to 0.

Theorem 1.2 Let $x = \left(x_{n,m,g}\right)$ be a sequence statistically convergent to L and $|x_{n,m,g} - L| \le M$ for every $(n,m,g) \in \mathbb{N}\times\mathbb{N}\times\mathbb{N}$. Then, it converges $N_{p,q}^{n,m,g}C_{n,m,g}^{(1,1,1)}$-statistically to L.

Proof: From the fact that $\left(x_{n,m,g}\right)$ converges statistically to L, we have

$$\lim_{n,m,g\to\infty}\frac{|i,j,k \le n,m,g : |x_{i,j,k} - L| \ge \varepsilon\}|}{nmg} = 0.$$

We will denote $B_\varepsilon = \{i,j,k \le n,m,g : |x_{i,j,k} - L| \ge \varepsilon\}$ and $\overline{B}_\varepsilon = \{i,j,k \le n,m,g : |x_{i,j,k} - L| \le \varepsilon\}$. Then,

$$|\frac{1}{R_{n,m,g}}\sum_{i=0}^{n}\sum_{j=0}^{m}\sum_{k=0}^{g}p_{i,j,k}q_{n-i,m-j,g-k}\frac{1}{(i+1)(j+1)(k+1)}\sum_{u=0}^{i}\sum_{v=0}^{j}\sum_{y=0}^{k}x_{u,v,y} - L|$$
$$= |\frac{1}{R_{n,m,g}}\sum_{i=0}^{n}\sum_{j=0}^{m}\sum_{k=0}^{g}p_{i,j,k}q_{n-i,m-j,g-k}\frac{1}{(i+1)(j+1)(k+1)}\sum_{u=0}^{i}\sum_{v=0}^{j}\sum_{y=0}^{k}(x_{u,v,y} - L)|$$

$$\leq \frac{1}{R_{n,m,g}} \sum_{\substack{i=0 \\ (i,j,k)\in B_\varepsilon}}^{n} \sum_{j=0}^{m} \sum_{k=0}^{g} p_{i,j,k} q_{n-i,m-j,g-k} \frac{1}{(i+1)(j+1)(k+1)} \sum_{u=0}^{i} \sum_{v=0}^{j} \sum_{y=0}^{k} |x_{u,v,y} - L|$$

$$+ \frac{1}{R_{n,m,g}} \sum_{\substack{i=0 \\ (i,j,k)\in \overline{B}_\varepsilon}}^{n} \sum_{j=0}^{m} \sum_{k=0}^{g} p_{i,j,k} q_{n-i,m-j,g-k} \frac{1}{(i+1)(j+1)(k+1)} \sum_{u=0}^{i} \sum_{v=0}^{j} \sum_{y=0}^{k} |x_{u,v,y} - L|$$

$$\leq M \frac{1}{R_{n,m,g}} \sum_{\substack{i=0 \\ (i,j,k)\in B_\varepsilon}}^{n} \sum_{j=0}^{m} \sum_{k=0}^{g} 1 + \varepsilon \leq M \frac{C_2}{nmg} \sum_{\substack{i=0 \\ (i,j,k)\in B_\varepsilon}}^{n} \sum_{j=0}^{m} \sum_{k=0}^{g} 1 + \varepsilon \to 0 + \varepsilon,$$

$$as \ n, m, g \to \infty,$$

for some constant C_2.

Converse of Theorem 1.2 is not true as can be seen in the following example.

Consider that $p_{n,m,g} = (n+1)(m+1)(g+1)$, $\left(q_{n,m,g}\right) = 1$ for some $(n,m,g) \in \mathbb{N}\cup\{0\} \times \mathbb{N}\cup\{0\} \times \mathbb{N}\cup\{0\}$ and define the sequence $x = \left(x_{n,m,g}\right)$ as follows:

$$x_{i,j,k} = \begin{cases} 1, & for \ i = p^2 - p, \ldots, p^2 - \ j = t^2 - t, \ldots, t^2 - 1 \ and \ k = o^2 - o, \ldots, o^2 - 1; \\ -\frac{1}{pto}, & for \ i - p^2, p - 2, .. \ j = t^2, t = 2, .. \ and \ k = o^2, o = 2, \ldots \\ 0, & otherwise \end{cases}$$

Under this conditions, after some basic calculations we get that $x = \left(x_{n,m,g}\right)$ is $N_{p,q}^{n,m,g} C_{n,m,g}^{(1,1,1)}$-summable to 1. Therefore, by Theorem 1.2, $x = \left(x_{n,m,g}\right)$ is $N_{p,q}^{n,m,g} C_{n,m,g}^{(1,1,1)}$-statistically convergent. On the other hand, the sequences $p^2; p = 2, 3, \ldots, t^2; t = 2, 3, \ldots$ and $o^2; o = 2, 3, \ldots$ have natural density zero and it is clear that $st\text{-}\liminf_{n,m,g} x_{n,m,g} = 0$ and $st\text{-}\limsup_{n,m,g} x_{n,m,g} = 1$. Hence, $\left(x_{i,j,k}\right)$ is not statistically convergent.

2. Tauberian theorems under $N_{p,q}^{n,m,g} C_{n,m,g}^{(1,1,1)}$-statistically convergence

In this section, we show the results that we obtained. Throughout this paper, $R_{\lambda_{n,m,g}}$ and $R_{\lambda_n,\lambda_m,\lambda_g}$ will have the same meaning.

Consider that $st\text{-}\lim i,j,k x_{i,j,k} = L$; $\left(x_{n,m,g}\right)$ is $N_{p,q}^{n,m,g} C_{n,m,g}^{(1,1,1)}$-statistically convergent and (13) satisfies, then for every $t > 1$, is valid the following relation

$$st - \lim_{i,j,k} \frac{1}{R_{\lambda_{i,j,k}} - R_{i,j,k}} \sum_{w=i+1}^{\lambda_i} \sum_{e=j+1}^{\lambda_j} \sum_{r=k+1}^{\lambda_k} p_{w,e,r} q_{\lambda_i - w, \lambda j - e, \lambda k - r} \frac{1}{(w+1)(e+1)(r+1)} \sum_{u=0}^{i} \sum_{v=0}^{j} \sum_{y=0}^{k} \left(x_{w,e,r} - x_{i,j,k}\right) = 0 \tag{7}$$

and in case where $0 < t < 1$,

$$st - \lim_{i,j,k} \frac{1}{R_{i,j,k} - R_{\lambda_{i,j,k}}} \sum_{w=\lambda_i+1}^{i} \sum_{e=\lambda_j+1}^{j} \sum_{r=\lambda_k+1}^{k} p_{w,e,r} q_{i-w,j-e,k-r}$$
$$\frac{1}{(w+1)(e+1)(r+1)} \sum_{u=0}^{i} \sum_{v=0}^{j} \sum_{y=0}^{k} (x_{i,j,k} - x_{w,e,r}) = 0. \tag{8}$$

The condition given by relation (13) is equivalent to the condition

$$st - \lim_{n,m,g \to \infty} \frac{R_{n,m,g}}{R_{\lambda_{n,m,g}}} > 1, \quad 0 < \lambda < 1. \tag{9}$$

Proof: Suppose that relation (13) is valid, $0 < \lambda < 1$, $w = \lambda_n = [\lambda n]$, $e = \lambda_m = [\lambda m]$ and $r = \lambda_g = [\lambda g]$, $(n,m,g) \in \mathbb{N} \times \mathbb{N} \times \mathbb{N}$. Then, it follows that

$$\frac{1}{\lambda} > 1 \Rightarrow \frac{w}{\lambda} = \frac{[\lambda n]}{t} \le n, \ \frac{1}{\lambda} > 1 \Rightarrow \frac{e}{\lambda} = \frac{[\lambda m]}{t} \le m \ \ and \ \ \frac{1}{\lambda} > 1 \Rightarrow \frac{r}{\lambda} = \frac{[\lambda g]}{t} \le g.$$

From above relation and definition of sequences $\left(p_{n,m,g}\right)$ and $\left(q_{n,m,g}\right)$, we have

$$\frac{R_{n,m,g}}{R_{\lambda_{n,m,g}}} \ge \frac{R_{\left[\frac{n}{\lambda}\right],\left[\frac{m}{\lambda}\right],\left[\frac{g}{\lambda}\right]}}{R_{\lambda_{n,m,g}}} \Rightarrow st - \liminf_{n,m,g \to \infty} \frac{R_{n,m,g}}{R_{\lambda_{n,m,g}}} \ge st - \liminf_{n,m,g \to \infty} \frac{R_{\left[\frac{n}{\lambda}\right],\left[\frac{m}{\lambda}\right],\left[\frac{g}{\lambda}\right]}}{R_{\lambda_{n,m,g}}} > 1.$$

Conversely, suppose that (9) is valid. Now, let $\lambda > 1$ be given and let $\lambda_1, \lambda_2, \lambda_3$ be chosen such that $1 < \lambda_1, \lambda_2, \lambda_3 < \lambda$. Set $w = \lambda_n = [\lambda n]$, $e = \lambda_m = [\lambda m]$ and $r = \lambda_g = [\lambda g]$. From $0 < \frac{1}{\lambda} < \frac{1}{\lambda_1}, \frac{1}{\lambda_2}, \frac{1}{\lambda_3} < 1$, it follows that

$$n \le \frac{\lambda n - 1}{\lambda_1} < \frac{[\lambda n]}{\lambda_1} = \frac{w}{\lambda_1}, \ m \le \frac{\lambda m - 1}{\lambda_2} < \frac{[\lambda m]}{\lambda_2} = \frac{e}{\lambda_2} \ \ and \ \ g \le \frac{\lambda g - 1}{\lambda_3} < \frac{[\lambda g]}{\lambda_3} = \frac{r}{\lambda_3}$$

provided $\lambda_1, \lambda_2, \lambda_3 \le \lambda - \frac{1}{n}, \lambda - \frac{1}{m}, \lambda - \frac{1}{g}$, which is a case where if n, m and g are large enough. Under this condition, we obtain

$$\frac{R_{\lambda_{n,m,g}}}{R_{n,m,g}} \ge \frac{R_{\lambda_{n,m,g}}}{R_{\left[\frac{w}{\lambda_1}\right],\left[\frac{e}{\lambda_2}\right],\left[\frac{r}{\lambda_3}\right]}} \Rightarrow st - \liminf_{n,m,g \to \infty} \frac{R_{\lambda_{n,m,g}}}{R_{n,m,g}} \ge st - \liminf_{n,m,g \to \infty} \frac{R_{\lambda_{n,m,g}}}{R_{\left[\frac{w}{\lambda_1}\right],\left[\frac{e}{\lambda_2}\right],\left[\frac{r}{\lambda_3}\right]}} > 1.$$

Consider that (13) is satisfied and let $x = \left(x_{i,j,k}\right)$ be a sequence of complex numbers which is $N_{p,q}^{n,m,g} C_{n,m,g}^{(1,1,1)}$-statistically convergent to L. Then,

$$st - \lim_{n,m,g} \frac{1}{R_{\lambda_{n,m,g}} - R_{n,m,g}} \sum_{i=n+1}^{\lambda_n} \sum_{j=m+1}^{\lambda_m} \sum_{k=g+1}^{\lambda_g} p_{i,j,k} q_{\lambda_n - i, \lambda_m - j, \lambda_g - k}$$
$$\frac{1}{(i+1)(j+1)(k+1)} \sum_{u=0}^{i} \sum_{v=0}^{j} \sum_{y=0}^{k} x_{u,v,y} = L \ for \ \lambda > 1 \tag{10}$$

and

$$st-\lim_{n,m,g}\frac{1}{R_{n,m,g}-R_{\lambda_{n,m,g}}}\sum_{i=\lambda_n+1}^{n}\sum_{j=\lambda_m+1}^{m}\sum_{k=\lambda_g+1}^{g}p_{i,j,k}q_{n-i,m-j,g-k}\frac{1}{(i+1)(j+1)(k+1)}\sum_{u=0}^{i}\sum_{v=0}^{j}\sum_{y=0}^{k}x_{u,v,y}=L\ for\ 0<\lambda<1. \tag{11}$$

Proof: We begin proving the case (10), i.e. when $\lambda>1$. Then, we have

$$\frac{1}{R_{\lambda_{n,m,g}}-R_{n,m,g}}\sum_{i=n+1}^{\lambda_n}\sum_{j=m+1}^{\lambda_m}\sum_{k=g+1}^{\lambda_g}p_{i,j,k}q_{\lambda_n-i,\lambda_m-j,\lambda_g-k}\frac{1}{(i+1)(j+1)(k+1)}\sum_{u=0}^{i}\sum_{v=0}^{j}\sum_{y=0}^{k}(x_{u,v,y}-L)$$
$$=\frac{R_{\lambda_{n,m,g}}}{R_{\lambda_{n,m,g}}-R_{n,m,g}}\frac{1}{R_{\lambda_{n,m,g}}}\sum_{i=n+1}^{\lambda_n}\sum_{j=m+1}^{\lambda_m}\sum_{k=g+1}^{\lambda_g}p_{i,j,k}q_{\lambda_n-i,\lambda_m-j,\lambda_g-k}\frac{1}{(i+1)(j+1)(k+1)}$$
$$\sum_{u=0}^{i}\sum_{v=0}^{j}\sum_{y=0}^{k}(x_{u,v,y}-L)-\frac{R_{n,m,g}}{R_{\lambda_{n,m,g}}-R_{n,m,g}}\frac{1}{R_{n,m,g}}\sum_{i=0}^{n}\sum_{j=0}^{m}\sum_{k=0}^{g}p_{i,j,k}q_{\lambda_n-i,\lambda_m-j,\lambda_g-k}$$
$$\frac{1}{(i+1)(j+1)(k+1)}\sum_{u=0}^{i}\sum_{v=0}^{j}\sum_{y=0}^{k}(x_{u,v,y}-L)$$
$$=\frac{R_{\lambda_{n,m,g}}}{R_{\lambda_{n,m,g}}-R_{n,m,g}}\frac{1}{R_{\lambda_{n,m,g}}}\sum_{i=n+1}^{\lambda_n}\sum_{j=m+1}^{\lambda_m}\sum_{k=g+1}^{\lambda_g}p_{i,j,k}q_{\lambda_n-i,\lambda_m-j,\lambda_g-j}$$
$$\frac{1}{(i+1)(j+1)(k+1)}\sum_{u=0}^{i}\sum_{v=0}^{j}\sum_{y=0}^{k}(x_{u,v,y}-L)-\frac{R_{n,m,g}}{R_{\lambda_{n,m,g}}-R_{n,m,g}}\frac{1}{R_{n,m,g}}\sum_{i=0}^{n}\sum_{j=0}^{m}\sum_{k=0}^{g}p_{i,j,k}$$
$$\left(q_{\lambda_n-i,\lambda_m-j,\lambda_g-k}+q_{n-i,m-j,g-k}-q_{n-i,m-j,g-k}\right)\frac{1}{(i+1)(j+1)(k+1)}\sum_{u=0}^{i}\sum_{v=0}^{j}\sum_{y=0}^{k}(x_{u,v,y}-L)$$
$$=\frac{R_{\lambda_{n,m,g}}}{R_{\lambda_{n,m,g}}-R_{n,m,g}}\frac{1}{R_{\lambda_{n,m,g}}}\sum_{i=n+1}^{\lambda_n}\sum_{j=m+1}^{\lambda_m}\sum_{k=g+1}^{\lambda_g}p_{i,j,k}q_{\lambda_n-i,\lambda_m-j,\lambda_g-k}$$
$$\frac{1}{(i+1)(j+1)(k+1)}\sum_{u=0}^{i}\sum_{v=0}^{j}\sum_{y=0}^{k}(x_{u,v,y}-L)-\frac{R_{n,m,g}}{R_{\lambda_{n,m,g}}-R_{n,m,g}}\frac{1}{R_{n,m,g}}\sum_{i=0}^{n}\sum_{j=0}^{m}\sum_{k=0}^{g}p_{i,j,k}$$
$$q_{n-i,m-j,g-k}\frac{1}{(i+1)(j+1)(k+1)}\sum_{u=0}^{i}\sum_{v=0}^{j}\sum_{y=0}^{k}x_{u,v,y}-\frac{R_{n,m,g}}{R_{\lambda_{n,m,g}}-R_{n,m,g}}\frac{1}{R_{n,m,g}}\sum_{i=0}^{n}\sum_{j=0}^{m}\sum_{k=0}^{g}p_{i,j,k}$$
$$\left(q_{\lambda_n-i,\lambda_m-j,\lambda_g-k}-q_{n-i,m-j,g-k}\right)\frac{1}{(i+1)(j+1)(k+1)}\sum_{u=0}^{i}\sum_{v=0}^{j}\sum_{y=0}^{k}(x_{u,v,y}-L). \tag{12}$$

From (12), definition of the sequence $\left(q_{n,m,g}\right)$ and relation $\limsup_{n,m,g} \frac{R_{\lambda_{n,m,g}}}{R_{\lambda_{n,m,g}} - R_{n,m,g}} < \infty$, we get (10).

Prove of (11) is made similarly to the prove of (10).

In the following theorem, we characterize the converse implication when the statistically convergence follows from its $N_{p,q}^{n,m,g} C_{n,m,g}^{(1,1,1)}$â^' statistically convergence.

Theorem 1.3 Let $\left(p_{n,m,g}\right)$ and $\left(q_{n,m,g}\right)$ be two non-negative real sequences and

$$st - \liminf_{n,m,g\to\infty} \frac{R_{\lambda_{n,m,g}}}{R_{n,m,g}} > 1 \; for \; every \; \lambda > 1, \tag{13}$$

where $\lambda_{n,m,g} = \lambda_n \lambda_m \lambda_g = [\lambda n][\lambda m][\lambda g]$ denotes the integral part of $\lambda n \lambda m \lambda g$ for every $(n,m,g) \in \mathbb{N} \times \mathbb{N} \times \mathbb{N}$, and let $\left(x_{n,m,g}\right)$ be a sequence of real numbers which is $N_{p,q}^{n,m,g} C_{n,m,g}^{(1,1,1)}$-statistically convergent to a finite number L. Then, $\left(x_{n,m,g}\right)$ is st-convergent to the same number L if and only if the following two conditions hold

$$\inf_{\lambda>1} \limsup_{n,m,g} \frac{1}{R_{n,m,g}} |\left\{ i,j,k \le R_{n,m,g} : \frac{1}{R_{\lambda_{i,j,k}} - R_{i,j,k}} \sum_{w=i+1}^{\lambda_i} \sum_{e=j+1}^{\lambda_j} \sum_{r=k+1}^{\lambda_k} p_{w,e,r} q_{\lambda_i - w, \lambda_j - e, \lambda_k - r} \right.$$
$$\left. \frac{1}{(w+1)(e+1)(r+1)} \sum_{u=0}^{i} \sum_{v=0}^{j} \sum_{y=0}^{k} \left(x_{w,e,r} - x_{i,j,k}\right) \le -\varepsilon \right\}| = 0, \tag{14}$$

and

$$\inf_{0<\lambda<1} \limsup_{n,m,g} \frac{1}{R_{n,m,g}} |\left\{ i,j,k \le R_{n,m,g} : \frac{1}{R_{i,j,k} - R_{\lambda_{i,j,k}}} \sum_{w=\lambda_i+1}^{i} \sum_{e=\lambda_j+1}^{j} \sum_{r=\lambda_k+1}^{k} p_{w,e,r} q_{i-w,j-e,k-r} \right.$$
$$\left. \frac{1}{(w+1)(e+1)(r+1)} \sum_{u=0}^{i} \sum_{v=0}^{j} \sum_{y=0}^{k} \left(x_{i,j,k} - x_{w,e,r}\right) \le -\varepsilon \right\}| = 0. \tag{15}$$

Proof: Necessity: Suppose that $\lim_{n,m,g\to\infty} x_{n,m,g} = L$ and (13) holds. By Proposition 2, we have

$$\lim_{n,m,g\to\infty} \frac{1}{R_{\lambda_{n,m,g}} - R_{n,m,g}} \sum_{i=n+1}^{\lambda_n} \sum_{j=m+1}^{\lambda_m} \sum_{k=g+1}^{\lambda_g} p_{i,j,k} q_{\lambda_n - i, \lambda_m - j, \lambda_g - k}$$
$$\frac{1}{(i+1)(j+1)(k+1)} \sum_{u=0}^{i} \sum_{v=0}^{j} \sum_{y=0}^{k} \left(x_{u,v,y} - x_{n,m,g}\right)$$
$$= \lim_{n,m,g\to\infty} \left\{ \left(\frac{1}{R_{\lambda_{n,m,g}} - R_{n,m,g}} \sum_{i=n+1}^{\lambda_n} \sum_{j=m+1}^{\lambda_m} \sum_{k=g+1}^{\lambda_g} p_{i,j,k} q_{\lambda_n - i, \lambda_m - j, \lambda_g - k} \right. \right.$$
$$\left. \left. \frac{1}{(i+1)(j+1)(k+1)} \sum_{u=0}^{i} \sum_{v=0}^{j} \sum_{y=0}^{k} x_{u,v,y} \right) - x_{n,m,g} \right\} = 0,$$

for every $\lambda > 1$. In case where $0 < \lambda < 1$, we have that

$$\lim_{n,m,g\to\infty} \frac{1}{R_{n,m,g} - R_{\lambda_{n,m,g}}} \sum_{i=\lambda_n+1}^{n} \sum_{j=\lambda_m+1}^{m} \sum_{k=\lambda_g+1}^{g} p_{i,j,k} q_{n-i,m-j,g-k}$$

$$\frac{1}{(i+1)(j+1)(k+1)} \sum_{u=0}^{i} \sum_{v=0}^{j} \sum_{y=0}^{k} \left(x_{n,m,g} - x_{u,v,y}\right)$$

$$= \lim_{n,m,g\to\infty} \left\{ x_{n,m,g} - \left(\frac{1}{R_{n,m,g} - R_{\lambda_{n,m,g}}} \sum_{i=\lambda_n+1}^{n} \sum_{j=\lambda_m+1}^{m} \sum_{k=\lambda_g+1}^{g} p_{i,j,k} q_{n-i,m-j,g-k} \right.\right.$$

$$\left.\left. \frac{1}{(i+1)(j+1)(k+1)} \sum_{u=0}^{i} \sum_{v=0}^{j} \sum_{y=0}^{k} x_{u,v,y} \right) \right\} - 0.$$

Sufficiency: Consider that (14) and ((15) are satisfied. In what follows, we will prove that $\lim_{n,m,g\to\infty} x_{n,m,g} = L$. Given any $\varepsilon > 0$, by (14) we can choose $\lambda_1 > 0$ such that

$$\liminf_{n,m,g\to\infty} \frac{1}{R_{\lambda_{n_1},\lambda_{m_1},\lambda_{g_1}} - R_{n,m,g}} \sum_{i=n+1}^{\lambda_{n_1}} \sum_{j=m+1}^{\lambda_{m_1}} \sum_{k=g+1}^{\lambda_{g_1}} p_{i,j,k} q_{\lambda_n-i,\lambda_m-j,\lambda_g-k}$$
$$\frac{1}{(i+1)(j+1)(k+1)} \sum_{u=0}^{i} \sum_{v=0}^{j} \sum_{y=0}^{k} \left(x_{u,v,y} - x_{n,m,g}\right) \geq -\varepsilon, \tag{16}$$

where $\lambda_{n_1} = [\lambda n_1]$, $\lambda_{m_1} = [\lambda m_1]$ and $\lambda_{g_1} = [\lambda g_1]$. By the assumed summability $N_{p,q}^{n,m,g} C_{n,m,g}^{(1,1,1)}$ of $(x_{n,m,g})$, Proposition 2 and (16), we have

$$\limsup_{n,m,g\to\infty} x_{n,m,g} \leq L + \varepsilon, \tag{17}$$

for any $\lambda > 1$.

On the other hand, if $0 < \lambda < 1$, for every $\varepsilon > 0$, we can choose $0 < \lambda_2 < 1$ such that

$$m \liminf_{n,m,g\to\infty} \frac{1}{R_{n,m,g} - R_{\lambda_{n_2},\lambda_{m_2},\lambda_{g_2}}} \sum_{i=\lambda_{n_2}+1}^{n} \sum_{j=\lambda_{m_2}+1}^{m} \sum_{k=\lambda_{g_2}+1}^{g} p_{i,j,k} q_{n-i,m-j,g-k}$$
$$\frac{1}{(i+1)(j+1)(k+1)} \sum_{u=0}^{i} \sum_{v=0}^{j} \sum_{y=0}^{k} \left(x_{n,m,g} - x_{u,v,y}\right) \geq -\varepsilon, \tag{18}$$

where $\lambda_{n_2} = [\lambda n_2]$, $\lambda_{m_2} = [\lambda m_2]$ and $\lambda_{g_2} = [\lambda g_2]$. By the assumed summability $N_{p,q}^{n,m,g} C_{n,m,g}^{(1,1,1)}$ of $(x_{n,m,g})$, Proposition 2 and (18), we have

$$\liminf_{n,m,g\to\infty} x_{n,m,g} \geq L - \varepsilon, \tag{19}$$

for any $0 < \lambda < 1$.

Since $\varepsilon > 0$ is arbitrary, combining (17) and (19), we obtain

$$\lim_{n,m,g\to\infty} x_{n,m,g} = L.$$

In the following theorem, we will consider the case where $x = \left(x_{n,m,g}\right)$ is a sequence of complex numbers.

Theorem 1.4 Let (13) be satisfied and let $\left(x_{n,m,g}\right)$ be a sequence of complex numbers which is $N_{p,q}^{n,m,g}C_{n,m,g}^{(1,1,1)}$-statistically convergent to a finite number L. Then, $\left(x_{n,m,g}\right)$ is convergent to the same number L if and only if the following two conditions hold

$$\inf_{\lambda>1} \limsup_{n,m,g} \frac{1}{R_{n,m,g}} |\left\{ i,j,k \le R_{n,m,g} : \frac{1}{R_{\lambda_{i,j,k}} - R_{i,j,k}} \sum_{w=i+1}^{\lambda_i} \sum_{e=j+1}^{\lambda_j} \sum_{r=k+1}^{\lambda_k} p_{w,e,r} q_{\lambda_i - w, \lambda_j - e, \lambda_k - r} \right.$$

$$\left. \frac{1}{(w+1)(e+1)(k+1)} \sum_{u=0}^{i} \sum_{v=0}^{j} \sum_{y=0}^{k} \left(x_{w,e,r} - x_{i,j,k}\right) \ge \varepsilon \right\}| = 0, \tag{20}$$

and

$$\inf_{0<\lambda<1} \limsup_{n,m,g} \frac{1}{R_{n,m,g}} |\left\{ i,j,k \le R_{n,m,g} : \frac{1}{R_{i,j,k} - R_{\lambda_{i,j,k}}} \sum_{w=\lambda_i+1}^{i} \sum_{e=\lambda_j+1}^{j} \sum_{r=\lambda_k+1}^{k} p_{w,e,r} q_{i-w,j-e,k-r} \right.$$

$$\left. \frac{1}{(w+1)(e+1)(r+1)} \sum_{u=0}^{i} \sum_{v=0}^{j} \sum_{y=0}^{k} \left(x_{i,j,k} - x_{w,e,r}\right) \ge \varepsilon \right\}| = 0. \tag{21}$$

Proof: Necessity: If both (2) and (6) hold, then Proposition 2 yields (20) for every $\lambda > 1$ and (21) for every $0 < \lambda < 1$.

Sufficiency: Suppose that (2), (13) and one of the conditions (20) and (21) are satisfied. For any given $\varepsilon > 0$, there exists $\lambda_1 > 0$ such that

$$\limsup_{n,m,g\to\infty} |\frac{1}{R_{\lambda_{n_1},\lambda_{m_1},\lambda_{g_1}} - R_{n,m,g}} \sum_{i=n+1}^{\lambda_{n_1}} \sum_{j=m+1}^{\lambda_{m_1}} \sum_{k=g+1}^{\lambda_{g_1}} p_{i,j,k} q_{\lambda_{n_1}-i,\lambda_{m_1}-j,\lambda_{g_1}-k}$$

$$\frac{1}{(i+1)(j+1)(k+1)} \sum_{u=0}^{i} \sum_{v=0}^{j} \sum_{y=0}^{k} \left(x_{u,v,y} - x_{n,m,g}\right)| \le \varepsilon,$$

where $\lambda_{n_1} = [\lambda n_1]$, $\lambda_{m_1} = [\lambda m_1]$ and $\lambda_{g_1} = [\lambda g_1]$. Taking into account fact that $\left(x_{n,m,g}\right)$ is $N_{p,q}^{n,m,g}C_{n,m,g}^{(1,1,1)}$ summbale to L and Proposition 2, we have the following estimation

$$\limsup_{n,m,g\to\infty} |L - x_{n,m,g}|$$

$$\leq \limsup_{n,m,g\to\infty} |L - \frac{1}{R_{\lambda_{n_1},\lambda_{m_1},\lambda_{g_1}} - R_{n,m}} \sum_{i=n+1}^{\lambda_{n_1}} \sum_{j=m+1}^{\lambda_{m_1}} \sum_{k=g+1}^{\lambda_{g_1}} p_{i,j,k} q_{\lambda_{n_1}-i,\lambda_{m_1}-j,\lambda_{g_1}-k}$$

$$\frac{1}{(i+1)(j+1)(k+1)} \sum_{u=0}^{i} \sum_{v=0}^{j} \sum_{y=0}^{k} x_{u,v,y}|$$

$$+\limsup_{n,m,g\to\infty} |L - \frac{1}{R_{\lambda_{n_1},\lambda_{m_1},\lambda_{g_1}} - R_{n,m}} \sum_{i=n+1}^{\lambda_{n_1}} \sum_{j=m+1}^{\lambda_{m_1}} \sum_{k=g+1}^{\lambda_{g_1}} p_{i,j,k} q_{\lambda_{n_1}-i,\lambda_{m_1}-j,\lambda_{g_1}-k}$$

$$\frac{1}{(i+1)(j+1)(k+1)} \sum_{u=0}^{i} \sum_{v=0}^{j} \sum_{y=0}^{k} (x_{u,v,y} - x_{n,m,g})|$$

$$\leq \varepsilon.$$

For a given $\varepsilon > 0$, there exists $\lambda_2 > 0$ such that

$$\limsup_{n,m,g\to\infty} | \frac{1}{R_{n,m,g} - R_{\lambda_{n_2},\lambda_{m_2},\lambda_{g_2}}} \sum_{i=\lambda_{n_2}+1}^{n} \sum_{j=\lambda_{m_2}+1}^{m} \sum_{k=\lambda_{g_2}+1}^{g} p_{i,j,k} q_{n-i,m-j,g-k}$$

$$\frac{1}{(i+1)(j+1)(k+1)} \sum_{u=0}^{i} \sum_{v=0}^{j} \sum_{y=0}^{k} (x_{n,m,g} - x_{u,v,y})| \leq \varepsilon,$$

where $\lambda_{n_2} = [\lambda n_2]$, $\lambda_{m_2} = [\lambda m_2]$ and $\lambda_{g_2} = [\lambda g_2]$. Taking into account fact that $(x_{n,m,g})$ is $N_{p,q}^{n,m,g} C_{n,m,g}^{(1,1,1)}$ summbale to L and Proposition 2, we obtain the following

$$\limsup_{n,m,g\to\infty} |L - x_{n,m,g}|$$

$$\limsup_{n,m\to\infty} |L - \frac{1}{R_{n,m,g} - R_{\lambda_{n_2},\lambda_{m_2},\lambda_{g_2}}} \sum_{i=\lambda_{n_2}+1}^{n} \sum_{j=\lambda_{m_2}+1}^{m} \sum_{k=\lambda_{g_2}+1}^{g} p_{i,j,k} q_{n-i,m-j,g-k}$$

$$\frac{1}{(i+1)(j+1)(k+1)} \sum_{u=0}^{i} \sum_{v=0}^{j} \sum_{y=0}^{k} x_{u,v,y}|$$

$$+\limsup_{n,m,g\to\infty} | \frac{1}{R_{n,m,g} - R_{\lambda_{n_2},\lambda_{m_2},\lambda_{g_2}}} \sum_{i=\lambda_{n_2}+1}^{n} \sum_{j=\lambda_{m_2}+1}^{m} \sum_{k=\lambda_{g_2}+1}^{g} p_{i,j,k} q_{n-i,m-j,g-k}$$

$$\frac{1}{(i+1)(j+1)(k+1)} \sum_{u=0}^{i} \sum_{v=0}^{j} \sum_{y=0}^{k} (x_{n,m,g} - x_{u,v,y})|$$

$$\leq \varepsilon.$$

Since $\varepsilon > 0$ in either case, we get

$$\lim_{n,m,g\to\infty} x_{n,m,g} = L.$$

3. Conclusion

In this paper, we have defined and proved new Tauberian theorems under triple statistically Nörlund-Cesáro summability, as a consequence of results showed in 2, some theorems, lemmas and corollaries can be defined and proved similarly by using $(1,0,0)$, $(0,1,0)$, and $(0,0,1)$ method of summability. It is well know that Tauberian theorems for single sequences of single variable have been achieved a high degree of development; however, it is still in its infancy for triple sequences. For that reason, the results established in this paper can be extended and studied in some inclusion, Tauberian type theorems and Tauberian convexity type for certain families of generalized Nörlund.

Author details

Carlos Granados
Universidad de Antioquia, Colombia

*Address all correspondence to: carlosgranadosortiz@outlook.es

References

[1] Fast H. Sur la convergence statistique. Colloquium Mathematicum. 1951;**2**: 241-244

[2] Steinhaus H. Sur la convergence ordinaire et la convergence asymptotique. Colloquium Mathematicum. 1951;**2**:73-74

[3] Fridy J. On statistical convergence. Analysis. 1985;**5**:301-313

[4] Jena BB, Paikray SK, Misra UK. Statistical deferred Cesaro summability and its applications to approximation theorems. Univerzitet u Nišu. 2018;**32**: 2307-2319

[5] Srivastava HM, Jena BB, Paikray SK, Misra UK. Deferred weighted Astatistical convergence based upon the (p, q)-Lagrange polynomials and its applications to approximation theorems. Journal of Applied Analysis. 2018;**24**: 1-16

[6] Parida P, Paikray SK, Jena B. Tauberian theorems for statistical Cesaro summability of function of two variables over a locally convex space. Studies in Computational Intelligence: Recent Advances in Intelligent Information Systems. 2020;**863**:779-790

[7] Tauber A. Ein satz der Theorie der unendlichen Reihen. Monatshefte für Mathematik. 1897;**8**:273-277

[8] Landau E. Uber einen Satz des Herrn Littlewood. Rendiconti del Circolo Matematico di Palermo. 1913;**35**:265-276

[9] Hardy GH, Littlewood JE. Tauberian theorems concerning power series and Dirichlets series whose coeffcients are positive. Proceedings of the London Mathematical Society. 1914;**13**:174-191

[10] Schmidt R. Uber divergente Folgen und lineare Mittelbildungen. Mathematische Zeitschrift. 1925;**22**: 89-152

[11] Braha N. Tauberian theorems under statistically Nörlund-Cesáro summability method. Journal of Mathematical Inequalities. 2020;**14**(4): 967-975

[12] Canak I, Totur U. Some Tauberian conditions for Cesaro summability method. Mathematica Slovaca. 2012;**62**: 271-280

[13] Jena BB, Paikray SK, Misra UK. A proof of Tauberian theorem for Cesaro summability method. Asian Journal of Mathematics and Computer Research. 2016;**8**:272-276

[14] Knopp K. Limitierungs-Umkehrsatze fur Doppelfolgen. Mathematische Zeitschrift. 1939;**45**: 573-589

[15] Moricz F. Tauberian theorems for Cesaro summable double sequences. Studia Mathematica. 1994;**110**:83-96

[16] Landau E. Uber die Bedeutung einiger neuerer Grenzwertsatze der Herren Hardy and Axer. Prace Mathematics FIZ. 1910;**21**:97-177

[17] Hardy GH. Divergent Series. Clarendon Press; 1949

[18] Canak I, Totur U. A Tauberian theorem for Cesaro summability of integrals. Applied Mathematics Letters. 2011;**24**:391-395

[19] Canak I, Totur U. Some classical Tauberian theorems for (C, 1, 1, 1, 1)

summable triple sequences. Georgian Mathematical Journal. 2016;**23**(1):33-42

[20] Canak I, Onder Z, Totur U. Statistical extensions of some classical Tauberian theorems for Cesaro summability of triple sequences. Results in Mathematics. 2016;**70**:457-473

[21] Totur U, Canak I. Some Tauberian conditions under which convergence follows from (C, 1, 1, 1) summability. Journal of Analysis. 2020;**28**:683-694

[22] Granados C, Das AK. New Tauberian theorems for statistical Cesaro summability of function of three variables over a locally convex space. Armenian Journal of Mathematics. 2022; **14**(5):1-15

[23] Granados C, Das AK, Das S. New Tauberian theorems for Cesaro summable triple sequences of fuzzy numbers. Kragujevac Journal of Mathematics. 2024;**48**(5):787-802

[24] Granados C. New notions of triple sequences on ideal spaces in metric spaces. Advances in the Theory of Nonlinear Analysis and its Applications. 2021;**5**(3):363-368

Chapter 3

A Brief Look at the Calderón and Hilbert Operators

Guillermo J. Flores

Abstract

The Calderón operator is the sum of the Hardy averaging operator and its adjoint, and plays an important role in the theory of real interpolation. On the other hand, the Hilbert operator arises from the continuous version of Hilbert's inequality. Both operators appear in different contexts and have numerous applications within harmonic analysis. In this chapter we will briefly review the Calderón and Hilbert operators, showing some of the most relevant results within functional analysis and finally we will present recent results on these operators within Fourier analysis.

Keywords: Calderón operator, Hilbert operator, Lebesgue spaces, Lipschitz spaces, BMO spaces, weighted inequalities, Calderón weights

1. Introduction

The Calderón and Hilbert operators are among the most relevant operators in harmonic analysis, arising from Hilbert's double series theorem which is one of the simplest and most beautiful in the theory of double series of positive terms. It was proved by Hilbert, in the course of his investigations in the theory of integral equations, that the series $\sum_{m,n \in \mathbb{N}} \frac{a_m a_n}{a_m + a_n}$, where $a_n \geq 0$ for all $n \in \mathbb{N}$, is convergent whenever $\sum_{n \in \mathbb{N}} a_n^2$ is convergent.

Other proofs of Hilbert's double series theorem and generalizations in different directions were studied and published over time by influential mathematicians such as H. Weyl, F. Wiener, J. Schur, Fejér and F. Riesz, Pólya and Szegö, Francis and Littlewood, G.H. Hardy and M. Riesz, among others.

In [1, 2], G.H. Hardy observed that Hilbert's theorem stated above is an immediate corollary of another theorem which has interest in itself. This theorem is as follows: If $a_n \geq 0$ for all $n \in \mathbb{N}$ and $\sum_{n \in \mathbb{N}} a_n^2$ is convergent, then $\sum_{n \in \mathbb{N}} \left(\frac{1}{n} \sum_{j=1}^{n} a_j \right)^2$ is also convergent.

The first extension of the just stated Hilbert's and Hardy's results in which we are interested is the following (see [3]): Let $1 < p < \infty$ and $p' = p/(p-1)$ (i.e. p' is the conjugate of p). If $\sum_{n=1}^{\infty} a_n^p$ and $\sum_{n=1}^{\infty} b_n^{p'}$ are convergent, where a_n and b_n are nonnegative numbers for all $n \in \mathbb{N}$, then

$$\sum_{m=1}^{\infty}\sum_{n=1}^{\infty}\frac{a_m b_n}{m+n} \le \frac{\pi}{\sin(\pi/p)}\left(\sum_{m=1}^{\infty}a_m^p\right)^{1/p}\left(\sum_{n=1}^{\infty}b_n^{p'}\right)^{1/p'} \quad \text{and} \sum_{n\in\mathbb{N}}\left(\frac{1}{n}\sum_{j=1}^{n}a_j\right)^p \le (p')^p\sum_{n=1}^{\infty}a_n^p.$$

The constants $\pi/\sin(\pi/p)$ and $(p')^p = (p/(p-1))^p$ are the best possible.

At the same time, the continuous versions of the previous inequalities are the following (see [3, 4]): Let $1<p<\infty$ and p' the conjugate of p. If $\int_{[0,\infty)}|f|^p$ and $\int_{[0,\infty)}|g|^{p'}$ are finite, then

$$\int_{[0,\infty)}\int_{[0,\infty)}\frac{|f(x)||g(y)|}{x+y}dx\,dy \le \frac{\pi}{\sin(\pi/p)}\left(\int_{[0,\infty)}|f(x)|^p dx\right)^{1/p}\left(\int_{[0,\infty)}|g(x)|^{p'}dx\right)^{1/p'}$$

and

$$\int_{[0,\infty)}\left(\frac{1}{x}\int_{[0,x]}f(y)dy\right)^p dx \le \left(\frac{p}{p-1}\right)^p\int_{[0,\infty)}|f(x)|^p dx.$$

Once again, the constants involved are the best possible.

As usual in harmonic analysis, if E is a measurable subset of $\mathbb{R}^n$, then $L^p(E)$, $1\le p<\infty$, is the Lebesgue space of all measurable functions f such that $\|f\|_{L^p(E)}^p = \int_E|f(x)|^p dx$ is finite. Recall that $\left(L^p(E), \|\cdot\|_{L^p(E)}\right)$ is a Banach space and in the case $E=\mathbb{R}^n$, it is denoted $\|\cdot\|_p = \|\cdot\|_{L^p(E)}$.

Now, consider the operators H and P defined by

$$Hf(x) = \int_{[0,\infty)}\frac{f(t)}{x+t}dt \qquad \text{and} \qquad Pf(x) = \frac{1}{x}\int_{[0,x]}f(t)dt,$$

which naturally arise from the inequalities presented above. Also consider

$$Qf(x) = \int_{[x,\infty)}\frac{f(t)}{t}dt$$

being the adjoint operator of P and satisfying

$$\int_{[0,\infty)}(Qf(x))^p dx = \int_{[0,\infty)}\left(\int_{[x,\infty)}\frac{f(t)}{t}dt\right)^p dx \le C\int_{[0,\infty)}(f(x))^p dx,$$

for all $f\in L^p([0,\infty))$, $1<p<\infty$, where C is a positive constant (see [4]). Therefore, P and Q are bounded operators from $L^p([0,\infty))$ in itself, that is,

$$\|Pf\|_{L^p([0,\infty))} \le C\|f\|_{L^p([0,\infty))} \text{ and } \|Qf\|_{L^p([0,\infty))} \le C\|f\|_{L^p([0,\infty))} \quad \text{for all } f\in L^p([0,\infty)).$$

It is immediate that for nonnegative functions f,

$$Hf(x) \le Pf(x) + Qf(x) \le 2Hf(x) \qquad \text{for all } x>0.$$

Consequently H is a bounded operator on $L^p([0,\infty))$, that is,

$$\|Hf\|_{L^p([0,\infty))} \leq C\|f\|_{L^p([0,\infty))} \qquad \text{for all } f \in L^p([0,\infty)).$$

It is well known that P is called the *Hardy averaging operator* and H is called the *Hilbert operator*. Also, the *Calderón operator* S is defined by $S = P + Q$, being then a bounded operator from $L^p([0,\infty))$ in itself.

We end this section with some of the first and most important direct applications obtained from Hilbert's and Hardy's inequalities.

Theorem 1.1 Let E be the interval $(0,1)$ and $f \in L^2(E)$ not null in E. Then

$$\sum_{n=0}^{\infty} \left(\int_E x^n f(x) dx \right)^2 < \pi \int_E f^2(x) dx$$

and the constant π is the best possible. The integrals $\int_E x^n f(x)dx$, $n = 0, 1, \ldots$ are called the *moments of f in E* and are important in many theories.

Theorem 1.2 (Carlema's inequalities) Let $\{a_n\}$ be a sequence of positive numbers and $1 < p < \infty$. Then

$$\sum_{n=1}^{\infty} \left(\frac{1}{n} \sum_{k=1}^{n} a_k{}^{1/p} \right)^p < \left(\frac{p}{p-1} \right)^p \sum_{n=1}^{\infty} a_n \quad \text{and} \quad \sum_{n=1}^{\infty} \left(\prod_{k=1}^{n} a_k \right)^{1/n} < e \sum_{n=1}^{\infty} a_n.$$

The constants involved are the best possible.

The corresponding integral version for the second inequality of Carlema's inequality is: If f is a positive function belonging to $L^1([0,\infty))$, then

$$\int_{[0,\infty)} \exp\left(\frac{1}{x} \int_{[0,x]} \log f(t) dt \right) dx = \int_{[0,\infty)} e^{P(\log f)(x)} dx < e \int_{[0,\infty)} f(x) dx.$$

where the constant e is the best possible.

Theorem 1.3 Let $1 < p \leq 2$ and p' the conjugate of p. If $Lf(s) = \int_0^\infty f(t) e^{-st} dt$, i.e. Lf is the Laplace transform of f, then

$$\int_0^\infty Lf(s)^{p'} ds \leq \frac{2\pi}{p'} \left(\int_0^\infty f(s)^p ds \right)^{p'/p} \qquad \text{for all } f \in L^p([0,\infty)).$$

Therefore L is a bounded operator from $L^p([0,\infty))$ into $L^{p'}([0,\infty))$, $1 < p \leq 2$, and $\|Lf\|_{p'} \leq (2\pi/p')^{1/p'} \|f\|_p$.

The number of applications and results that arise from Hilbert's and Hardy's inequalities is by now very large and it would be impossible to give a detailed survey of all of them in a reasonable amount of text. We have simply made a very brief introduction about them in this section.

2. Calderón weights and L^p-weighted inequalities

A function ω defined on $\mathbb{R}^n$ is called a *weight* if it is locally integrable and positive almost everywhere. For a measurable set $E \subset \mathbb{R}^n$, $|E|$ denote its Lebesgue measure, $\omega(E) = \int_E \omega$, and E^c the complement of E in $\mathbb{R}^n$. Given a ball B, tB is the ball with the same center as B and with radius t times as long, and $f_B = \frac{1}{|B|}\int_B f$. As usual, χ_E denotes the characteristic function of E and $B(x, r)$ denotes a ball centered at x with radius r. Also, C denotes a positive constant.

Let ω be a weight in $\mathbb{R}^n$ and $1 \le p < \infty$. A Lebesgue measurable function f belongs to $L^p(\omega)$ if

$$\|f\|_{L^p(\omega)} = \left(\int_{\mathbb{R}^n} |f|^p \omega\right)^{1/p} < \infty.$$

We say that an oprator T is a bounded operator on $L^p(\omega)$ if

$$\|Tf\|_{L^p(\omega)} \le C\|f\|_{L^p(\omega)}, \qquad \text{for all} f \in L^p(\omega).$$

Given $1 < p < \infty$, it is said that ω is a Calderón weight of class C_p, that is $\omega \in C_p$, if the Calderón operator S is bounded on $L^p(\omega)$ (see [5]) or, equivalently, if P and Q are both bounded on $L^p(\omega)$ (see also [6]). It is well known that the class C_p for $p > 1$ is given by the conditions

$$M_p: \quad \left(\int_{[0,x]} \omega(t)dt\right)^{1/p} \left(\int_{[x,\infty)} \frac{\omega^{1-p'}(t)}{t^{p'}}dt\right)^{1/p'} \le C \quad \text{for all } x > 0;$$

$$M^p: \quad \left(\left(\int_{[x,\infty)} \frac{\omega(t)}{t^p}dt\right)^{1/p} \left(\int_{[0,x]} \omega^{1-p'}(t)dt\right)^{1/p'} \le C \quad \text{for all } x > 0.$$

The Calderón operator plays an important role in the theory of real interpolation and such theory related to Calderón weights is developed in [5]. On the other hand, in [7], the authors considered a maximal operator N on $(0, \infty)$ associated to the basis of open sets of the form $(0, b)$, given by

$$Nf(x) = \sup_{b > x} \frac{1}{b} \int_{[0,b]} |f(t)|dt$$

for measurable functions f. Then, for nonnegative functions f, we have

$$P(x) \le Nf(x) \le Sf(x) \qquad \text{for all } x > 0.$$

The classes of weights ω associated to the boundedness of N on $L^p(\omega)$ are those that satisfy the Muckenhoupt-A_p condition, $1 \le p < \infty$, only for the sets of the form $(0, b)$. These classes are denoted by $A_{p,0}$ and defined as follows:

$$A_{1,0}: \quad N\omega(x) \le C\omega(x) \quad \text{for almost all } x > 0;$$

$$A_{p,0}: \quad \left(\frac{1}{x}\int_{[0,x]} \omega\right)\left(\frac{1}{x}\int_{[0,x]} \omega^{1-p'}\right)^{p-1} \le C \quad \text{for all } x > 0, \text{where} 1 < p < \infty.$$

Then, in [7] is proved that N and S are bounded operators on $L^p(\omega)$ if and only if $\omega \in A_{p,0}$ for $1 < p < \infty$. This result implies, in particular, that the classes of weights C_p and $A_{p,0}$ coincide for $1 < p < \infty$.

Taking into account these results it is natural to wonder for the action of the Calderón and Hilbert operators over suitable spaces such as *BMO* or Lipschitz spaces. Also, another interesting question is: which are, in these cases, the Calderón weights in order to obtain weighted inequalities between these spaces?

These problems were treated for instance in the case of the fractional integral operator in [8, 9], which have been the main motivation for the article [10] and for the development of the following sections.

3. The n-dimensional Calderón and Hilbert operators

For $0 \leq \alpha < n, f$ a Lebesgue measurable function and $x \in \mathbb{R}^n$, $x \neq 0$, the general n-dimensional Calderón and Hilbert operators are defined by

$$S_\alpha f(x) = P_\alpha f(x) + Q_\alpha f(x) \quad \text{and} \quad H_\alpha f(x) = \int_{\mathbb{R}^n} \frac{f(y)}{(|x| + |y|)^{n-\alpha}} dy,$$

where $P_\alpha f(x) = \frac{1}{|x|^{n-\alpha}} \int_{|y| \leq |x|} f(y) dy$ and $Q_\alpha f(x) = \int_{|y| > |x|} \frac{f(y)}{|y|^{n-\alpha}} dy$.

Again, it is immediate that for nonnegative functions f, the following pointwise inequalities hold

$$H_\alpha f(x) \leq S_\alpha f(x) \leq 2^{n-\alpha} H_\alpha f(x), \tag{1}$$

and consequently, all weighted-L^p inequalities obtained for S are true for H and reciprocally.

In spite of the punctual comparison (1), we will show in Section 4 that the results obtained for S_α and H_α are not analogous when the BMO^γ and Lipschitz spaces are involved.

Both operators S_α and H_α appear in several different contexts and applications, see for instance [4, 11–17].

Next, we introduce the spaces of functions and the classes of weights which appear in our main results.

Recall that a measurable function f defined on $E \subset \mathbb{R}^n$ is said to be *essentially bounded* provided there is some $M \geq 0$, called an *essential upper bound* for f, for which $|f(x)| \leq M$ for almost all $x \in E$. As usual, the class of all functions that are essentially bounded on E is denoted by $L^\infty(E)$ and $\|f\|_\infty$ is the infimum of the essential upper bounds for $f \in L^\infty(E)$. Then, $(L^\infty(E), \|\cdot\|_\infty)$ is a Banach space.

Now, a Lebesgue measurable function f belongs to $L^\infty(\omega)$ if $\|f\omega\|_\infty < \infty$.

Also recall that $L^1_{loc}(\mathbb{R}^n)$ denotes the space of locally integrable functions f satisfying that $\|f\chi_B\|_1$ is finite for every ball $B \subset \mathbb{R}^n$.

Definition 3.2. Let ω be a weight in $\mathbb{R}^n$ and $0 \leq \gamma < 1/n$. A locally integrable function f belongs to $BMO^\gamma(\omega)$ if there exists a constant C such that for every ball $B \subset \mathbb{R}^n$,

$$\frac{1}{\omega(B)|B|^\gamma} \int_B |f - f_B| \leq C. \tag{2}$$

The seminorm of $f \in BMO^\gamma(\omega)$, $\|f\|_{BMO^\gamma(\omega)}$, is the infimum of all such C.

Definition 3.4. Let ω be a weight in $\mathbb{R}^n$ and $0 \le \gamma < 1/n$. A locally integrable function f belongs to $BM_0^\gamma(\omega)$ if there exists a constant C such that

$$\frac{1}{\omega(B)|B|^\gamma}\int_B |f| \le C \tag{3}$$

for every ball $B \subset \mathbb{R}^n$ centered at the origin.

The norm of $f \in BM_0^\gamma(\omega)$, denoted by $\|f\|_{BM_0^\gamma(\omega)}$, is the infimum of all such C. We will denote by $BM_0(\omega) = BM_0^0(\omega)$.

Observe that with these definitions the space $BMO^0(\omega)$ is the weighted version of BMO introduced by Muckenhoupt and Wheeden in [18]. Also, the family of spaces $BMO^\gamma(\omega)$ is contained in the family of weighted Lipschitz spaces $\mathcal{I}_\omega(\gamma)$ defined and studied in [8], and $BMO^\gamma(\omega)$ for $\omega \equiv 1$ is the well known Lipschitz integral space. Furthermore, we note that $L^\infty(\omega^{-1}) \subset BM_0(\omega) \cap BMO(\omega)$.

Given $p > 1$, it is known that a weight ω satisfies the reverse Hölder inequality with exponent p, denoted by $\omega \in RH(p)$, if

$$\left(\frac{1}{|B|}\int_B \omega^p\right)^{1/p} \le C\frac{1}{|B|}\int_B \omega \tag{4}$$

for all balls $B \subset \mathbb{R}^n$ and some constant C.

Definition 3.7. Given $p > 1$, a weight ω belongs to $RH_0(p)$ if it satisfies (4) but only for balls centered at the origin.

Definition 3.8. A weight ω belongs to D_0 if it satisfies the doubling condition $\omega(2B) \le C\omega(B)$ for every ball $B \subset \mathbb{R}^n$ centered at the origin and some constant C.

Definition 3.9. Let $\eta \ge 1$, a weight ω belongs to D_η if it satisfies the doubling condition

$$\frac{\omega(2B(x, |x|+r))}{|B(x, |x|+r)|^\eta} \le C\frac{\omega(B(x,r))}{|B(x,r)|^\eta}$$

every ball $B(x,r) \subset \mathbb{R}^n$ and some constant C.

It is immediate that $D_\eta \subset D_0$ for all η, and D_η is increasing in η. It is well known that each weight in the Muckenhoupt class A_∞ is in $RH(p) \cap D_\eta$ for some p and for some η, see for instance [19]. On the other hand, there exist weights belonging to D_η for some η, such that it is not in A_∞, see [20].

Also, we observe the following property that we will use along this chapter. If $\omega \in D_\eta$ there exists a constant C such that

$$\omega(B) \le C\omega\left(B\backslash\frac{1}{2}B\right) \tag{5}$$

for every ball $B \subset \mathbb{R}^n$ centered at the origin.

Definition 3.11. Let $0 \le \alpha < n$ and $1 < p < \infty$. A weight ω belongs to $\mathrm{H}_0(\alpha, p)$ if there exists a constant C such that

$$\left(\int_{B^c}\frac{\omega^{p'}(y)}{|y|^{(n-\alpha+1)}p'}dy\right)^{1/p'} \le C\frac{\omega(B)}{|B|^{1+1/p-\alpha/n+1/n}} \tag{6}$$

for every ball $B \subset \mathbb{R}^n$ centered at the origin.

A weight ω belongs to $\mathrm{H}_0(\alpha, \infty)$ if there exists a constant C such that

$$\int_{B^c} \frac{\omega(y)}{|y|^{n-\alpha+1}} dy \leq C \frac{\omega(B)}{|B|^{1-\alpha/n+1/n}} \tag{7}$$

for every ball $B \subset \mathbb{R}^n$ centered at the origin.

The classes of weights $\mathrm{H}_0(\alpha, p)$ and $\mathrm{H}_0(\alpha, \infty)$ satisfying (6) and (7) respectively but for all ball $B \subset \mathbb{R}^n$, were introduced and studied in [8].

4. Weighted Lebesgue and BMO^γ norm inequalities for S_α and H_α

Before beginning our study of the generalized Calderón operator, we notice that $S_\alpha f$ can be identically infinite for some functions f belonging to $L^p(\omega^{-p})$ or $BM_0^\gamma(\omega)$. For example, for $\omega \equiv 1$ and $\alpha > 0$, if $f(x) = |x|^{-\alpha}\chi_{B^c(0,1)}(x)$ and $n/\alpha < p$, then $f \in L^p(\omega^{-p})$ but $S_\alpha f \equiv \infty$. For the case $n/\alpha = p$, if $g(x) = |x|^{-\alpha}(\log|x|)^{-(1+1/p)/2}\chi_{B^c(0,2)}(x)$, then $g \in L^p(\omega^{-p})$ but $S_\alpha g \equiv \infty$. Also, if $h(x) = \chi_{B^c(0,1)}(x)$, then $h \in BM_0^\gamma(\omega)$ but $S_\alpha h \equiv \infty$ for all $0 \leq \alpha < n$. However, in Lemma 4.7 we will show that if f belongs to $L^p(\omega^{-p}) \cup BM_0^\gamma(\omega)$ and $S_\alpha f(x)$ is finite for some $x \neq 0$, then $S_\alpha f$ is finite on $\mathbb{R}^n \setminus \{0\}$. This also happens for the generalized Hilbert operator since the comparison (1).

Therefore, throughout the following sections we shall consider S_α and H_α defined on functions f belonging to $L^p(\omega^{-p})$ or $BM_0^\gamma(\omega)$ such that $S_\alpha f$ and $H_\alpha f$ are finite for some $x \neq 0$.

Also, note that $S_\alpha f$ is finite on $\mathbb{R}^n \setminus \{0\}$ for all compactly supported functions $f \in L^\infty(\omega^{-1})$, and the same holds for $H_\alpha f$. These functions belongs to $L^p(\omega^{-p})$ and those such that zero is not in their support belongs to $BM_0^\gamma(\omega)$.

The operator P is naturally bounded from BM_0 into L^∞ and analogously, Q is naturally bounded from BM_0 into BMO (see Proposition 3.5 in [13]). So, immediately the Calderón operator is bounded from BM_0 into BMO. This natural boundedness is our motivation in order to consider the $BM_0^\gamma(\omega)$ spaces and obtain Theorems 1.5 and 1.7. Likewise, since $L^\infty(\omega^{-1}) \subset BM_0(\omega)$, we get Corollaries 4.1 and 4.2.

We now state the main results of this chapter.

Theorem 1.4 Suppose $\alpha > 0$, $n/\alpha \leq p < n/(\alpha-1)^+$, $\eta = 1 + 1/n + 1/p - \alpha/n$ and $\delta = \alpha/n - 1/p$. The operator S_α is bounded from $L^p(\omega^{-p})$ into $BMO^\delta(\omega)$ and $\omega^{p'} \in D_0$ if and only if $\omega \in RH_0(p') \cap D_\eta$.

Theorem 1.5 Suppose $0 \leq \alpha < 1$, $0 \leq \gamma < 1/n - \alpha/n$, $\eta = 1 + 1/n - \alpha/n - \gamma$ and $\delta = \alpha/n + \gamma$. The operator S_α is bounded from $BM_0^\gamma(\omega)$ into $BMO^\delta(\omega)$ and $\omega \in D_0$ if and only if $\omega \in D_\eta$.

Corollary 4.1. *Let $\eta = 1 + 1/n$. Then S is bounded from $L^\infty(\omega^{-1})$ into $BMO(\omega)$ and $\omega \in D_0$ if and only if $\omega \in D_\eta$.*

Theorem 1.6 Suppose $\alpha > 0$, $n/\alpha \leq p < n/(\alpha-1)^+$, $\eta = 1 + 1/n + 1/p - \alpha/n$ and $\delta = \alpha/n - 1/p$. The operator H_α is bounded from $L^p(\omega^{-p})$ into $BMO^\delta(\omega)$ if and only if $\omega \in \mathrm{H}_0(\alpha, p) \cap RH_0(p') \cap D_\eta$.

Theorem 1.7 Suppose $0 \leq \alpha < 1$, $0 \leq \gamma < 1/n - \alpha/n$, $\eta = 1 + 1/n - \alpha/n - \gamma$ and $\delta = \alpha/n + \gamma$. The operator H_α is bounded from $BM_0^\gamma(\omega)$ into $BMO^\delta(\omega)$ if and only if $\omega \in \mathrm{H}_0(\alpha + n\gamma, \infty) \cap D_\eta$.

Corollary 4.2. *Let $\eta = 1 + 1/n$. Then H is bounded from $L^\infty(\omega^{-1})$ into $BMO(\omega)$ if and only if $\omega \in \mathrm{H}_0(0,\infty) \cap D_\eta$.*

Remark 4.3. It is classic the study of the boundedness of operators between L^∞ and BMO spaces. In [10], the results obtained in Corollaries 4.1 and 4.2 are originals, even in the unweighted case for H. The unweighted case for S is contained in Proposition 3.5 of [13].

Remark 4.4. The limit case $p = \infty$ $(p' = 1)$ of Theorem 1.4 is contained in Theorem 1.5 with $\gamma = 0$, since the hypotheses on the weights coincide. This also is true to Theorems 1.6 and 1.7.

Let α, p and η be as in Theorems 1.4 and 1.6. It is not difficult to show that if $\omega^{p'} \in A_{1,0}$ then $\omega \in \mathrm{H}_0(\alpha,p) \cap RH_0(p') \cap D_\eta$. Also, if $\omega(x) = |x|^\beta$ with $\beta \in (0, 1 + n/p - \alpha)$, then $\omega^{p'} \notin A_{1,0}$ but $\omega \in \mathrm{H}_0(\alpha,p) \cap RH_0(p') \cap D_\eta$. Furthermore, if $\omega(x) = |x|^\beta$ with $\beta = 1 + n/p - \alpha$, then $\omega \in RH_0(p') \cap D_\eta$ but $\omega \notin \mathrm{H}_0(\alpha,p)$. Now, if in addition $0 < \alpha < 1$ and $p' > n/(1-\alpha)$, we have that if $\omega^{p'} \in A_{p'+1,0}$ then $\omega \in \mathrm{H}_0(\alpha,p) \cap RH_0(p') \cap D_\eta$. In fact, the $\mathrm{H}_0(\alpha,p)$-condition is obtained directly from the $A_{p'+1,0}$-condition, and by Hölder inequality we have that

$$\left(\frac{\omega^{p'}(B(0,2(|x_0|+r)))}{|B(0,2(|x_0|+r))|}\right)^{1/p'} \leq C\frac{|B(0,2(|x_0|+r))|}{\omega^{-1}(B(x_0,r))} \leq C\left(\frac{|x_0|+r}{r}\right)^n \frac{\omega(B(x_0,r))}{|B(x_0,r)|}$$

$$\leq C\left(\frac{|x_0|+r}{r}\right)^{1-\alpha+n/p} \frac{\omega(B(x_0,r))}{|B(x_0,r)|}$$

for all balls $B(x_0,r) \subset \mathbb{R}^n$. Thus, the $RH_0(p')$ and D_η conditions follow from the last expression.

On the other hand, suppose that α, γ and η be as in Theorems 1.5 and 1.7. If $\omega \in A_{1,0}$ then $\omega \in \mathrm{H}_0(\alpha + n\gamma, \infty) \cap D_\eta$. Also, if $\omega(x) = |x|^\beta$ with $\beta \in (0, 1 - \alpha - n\gamma)$, then $\omega \notin A_{1,0}$ but $\omega \in \mathrm{H}_0(\alpha + n\gamma, \infty) \cap D_\eta$. Finally, if $\omega(x) = |x|^\beta$ with $\beta = 1 - \alpha - n\gamma$, then $\omega \in D_\eta$ but $\omega \notin \mathrm{H}_0(\alpha + n\gamma, \infty)$.

We shall denote by $A(x,r,R)$ with $0 < r < R$ the annulus centered at x with radii r and R, and by C and c positive constants not necessarily the same at each occurrence.

Before proceeding to the proofs of the main theorems we give some previous lemmas.

Suppose that $1 < p < \infty$ and $\omega \in RH_0(p')$, then it is easy to see that there exists C such that

$$\int_B |f| \leq C\frac{\omega(B)}{|B|^1/p}\|f\|_{L^p(\omega^{-p})} \tag{8}$$

for all $f \in L^p(\omega^{-p})$ and for every ball $B \subset \mathbb{R}^n$ centered at the origin.

Lemma 4.6. *(i) Let $0 < \alpha < n$ and $1 < p < \infty$. If $\omega \in \mathrm{H}_0(\alpha,p)$ then there exists C such that*

$$\int_{B^c} \frac{|f(y)|}{|y|^{n-\alpha+1}} dy \leq C\frac{\omega(B)}{|B|^{1+1/p-\alpha/n+1/n}}\|f\|_{L^p(\omega^{-p})}$$

for all $f \in L^p(\omega^{-p})$ and for every ball $B \subset \mathbb{R}^n$ centered at the origin.

(ii) Let $0 \leq \alpha < 1$, $0 \leq \gamma < 1/n - \alpha/n$ and $\eta = 1 + 1/n - \alpha/n - \gamma$. If $\omega \in \mathrm{H}_0(\alpha + n\gamma, \infty) \cap D_\eta$ then there exists C such that

$$\int_{B^c} \frac{|f(y)|}{|y|^{n-\alpha+1}} dy \leq C \frac{\omega(B)}{|B|^{\eta}} \|f\|_{BM_0^{\gamma}(\omega)}$$

for all $f \in BM_0^{\gamma}(\omega)$ and for every ball $B \subset \mathbb{R}^n$ centered at the origin.

Proof: The part (i) is immediate from Hölder's inequality and Definition 3.11. For (ii), since the hypothesis on ω and (3.10), for $B = B(0,r)$ we have

$$\begin{aligned}
\int_{B^c} \frac{|f(y)|}{|y|^{n-\alpha+1}} dy &\leq C \sum_{k=0}^{\infty} \frac{1}{(2^k r)^{n-\alpha+1}} \int_{2^k r \leq |y| < 2^{k+1} r} |f(y)| dy \\
&\leq C \|f\|_{BM_0^{\gamma}(\omega)} \sum_{k=0}^{\infty} \frac{\omega(B(0, 2^{k+1} r))}{(2^k r)^{n-\alpha+1-n\gamma}} \\
&\leq C \|f\|_{BM_0^{\gamma}(\omega)} \sum_{k=0}^{\infty} \frac{\omega(B(0, 2^{k+1} r) \backslash B(0, 2^k r))}{(2^k r)^{n-(\alpha+n\gamma)+1}} \\
&\leq C \|f\|_{BM_0^{\gamma}(\omega)} \frac{\omega(B)}{|B|^{\eta}}.
\end{aligned}$$

Lemma 4.7. *(i) Let $\alpha > 0$, $1 < p < \infty$ and $\omega \in RH_0(p')$. If $f \in L^p(\omega^{-p})$ and there exists $x \neq 0$ such that $S_\alpha f(x)$ is finite, then $S_\alpha f$ is finite on $\mathbb{R}^n \backslash \{0\}$ and $S_\alpha f \in L^1_{loc}(\mathbb{R}^n)$. The claim also holds for H_α.*

(ii) Let $\omega \in D_\eta$. If $f \in BM_0^{\gamma}(\omega)$ and there exists $x \neq 0$ such that $S_\alpha f(x)$ is finite, then $S_\alpha f$ is finite on $\mathbb{R}^n \backslash \{0\}$ and $S_\alpha f \in L^1_{loc}(\mathbb{R}^n)$. The claim also holds for H_α.

Proof: Since (3.1) we will only consider the operator S_α. Suppose f is a nonnegative function in $L^1_{loc}(\mathbb{R}^n)$ such that $S_\alpha f(x_0) < \infty$ for some $x_0 \neq 0$. Then $Q_\alpha f(x) < \infty$ for $|x| \geq |x_0|$, and if $0 < |x| < |x_0|$ then

$$Q_\alpha f(x) \leq \frac{1}{|x|^{n-\alpha}} \int_{|x| < |y| < |x_0|} f(y) dy + Q_\alpha f(x_0) < \infty.$$

Furthermore, since

$$\int_{B(0,r)} (Q_\alpha f(x) - Q_\alpha f(\nu)) dx \leq \int_{B(0,r)} f(y) r^{\alpha} dy < \infty,$$

where $|\nu| = r$, then $Q_\alpha f \in L^1_{loc}(\mathbb{R}^n)$.

If $\alpha > 0$ it is immediate that $P_\alpha f \in L^1_{loc}(\mathbb{R}^n)$. Therefore, (i) follows from (4.5). For (ii) it remains to show that $P_\alpha f \in L^1_{loc}(\mathbb{R}^n)$ in the case $\alpha = 0$. Let $B_j = B(0, 2^{-j} r), j = 0, 1, \ldots,$ by (3.10) we have

$$\begin{aligned}
\int_{B_0} \frac{1}{|x|^n} \int_{B(0,|x|)} f(y) dy dx &\leq C \|f\|_{BM_0^{\gamma}(\omega)} \int_{B_0} \frac{\omega(B(0,|x|))}{|x|^{n-n\gamma}} dx \\
&\leq C \|f\|_{BM_0^{\gamma}(\omega)} \sum_{j=0}^{\infty} \frac{r^{n\gamma-n}}{2^{j(n\gamma-n)}} \int_{B_j \backslash B_{j+1}} \omega(B_j) dx \\
&\leq C \|f\|_{BM_0^{\gamma}(\omega)} r^{n\gamma} \sum_{j=0}^{\infty} \frac{\omega(B_j \backslash B_{j+1})}{2^{jn\gamma}} \\
&\leq C \|f\|_{BM_0^{\gamma}(\omega)} r^{n\gamma} \omega(B_0).
\end{aligned}$$

Proof of Theorem 1.4: We begin showing the sufficient condition. Let $B = B(x_0, r)$. If $x_0 = 0$, let $u = re_1/2$ and $v = 3re_1/4$, where $e_1 = (1, \dots, 0)$. If $x_0 \neq 0$, let $u = (|x_0|+r/2)x_0/|x_0|$ and $v = (|x_0|+3r/4)x_0/|x_0|$. Thus, we consider the following two regions

$$\begin{aligned} U &= B(u, r/8) \cap \{u + h : \text{sign}(u_i) = \text{sign}(h_i) \ i = 1, \dots, n\}, \\ V &= B(v, r/4) \cap \{v + h : \text{sign}(v_i) = \text{sign}(h_i) \ i = 1, \dots, n\}, \end{aligned} \tag{9}$$

where $u = (u_1, \dots, u_n)$, $v = (v_1, \dots, v_n)$ and $h = (h_1, \dots, h_n)$. In the case $u_i = 0$ for some i, we choose $h_i > 0$. Clearly, we have the estimates $\text{dist}(U, V) = Cr$,

$$|U| = \frac{1}{2^n}|B(u, r/8)| = C|B| \quad \text{and} \quad |V| = \frac{1}{2^n}|B(v, r/4)| = C|B|.$$

Let f a nonnegative function in $L^p(\omega^{-p})$ such that $\text{supp}(f) \subset B(0, |x_0|+r/2)$, where $\text{supp}(f)$ is the closure of the set $\{x : f(x) \neq 0\}$. Then

$$\begin{aligned} \|S_\alpha f\|_{BMO^\delta(\omega)} &\geq \frac{C}{\omega(B)|B|^{1+\delta}} \int_B \int_B |S_\alpha f(x) - S_\alpha f(z)| dz dx \\ &\geq \frac{C}{\omega(B)|B|^{1+\delta}} \int_U \int_V \left| \left(\frac{1}{|x|^{n-\alpha}} - \frac{1}{|z|^{n-\alpha}} \right) \int_{B(0,|x_0|+r/2)} f(y) dy \right| dz dx. \end{aligned}$$

Note that, for $x \in U$ and $z \in V$ we have $\frac{1}{|x|^{n-\alpha}} - \frac{1}{|z|^{n-\alpha}} \geq C \frac{r}{(|x_0|+r)^{n-\alpha+1}}$. Then

$$\|S_\alpha f\|_{BMO^\delta(\omega)} \geq \frac{Cr^{n+1}}{\omega(B)|B|^\delta (|x_0|+r)^{n-\alpha+1}} \int_{B(0,|x_0|+r/2)} f(y) dy. \tag{10}$$

Thus, taking $f(y) = \omega^{p'}(y)\chi_{B(0,|x_0|+r/2)}(y)$ in (10) and since the boundedness of S_α and $\omega^{p'} \in D_0$, we have

$$\left(\frac{\omega^{p'}(B(0, |x_0|+r)}{|B(0, |x_0|+r)|} \right)^{1/p'} \leq C \left(\frac{|x_0| + r}{r} \right)^{1-\alpha+n/p} \frac{\omega(B)}{|B|}.$$

Taking $x_0 = 0$ in the last expression, we have that $\omega \in RH_0(p')$. Then, applying the Hölder's inequality, we obtain that ω satisfies the desired condition D_η.

Now, let us show the necessary condition. Let $f \in L^p(\omega^{-p})$ such that $S_\alpha f(x)$ is finite for some $x \neq 0$ and let $\omega \in RH_0(p') \cap D_\eta$. It is immediate that $\omega^{p'} \in D_0$. Thus $S_\alpha f \in L^1_{loc}(\mathbb{R}^n)$ by (i) of Lemma 4.7. First, we consider $B = B(0, r)$, $x \in B$ and $x \neq 0$. Let ν be such that $|\nu| = r$, and let

$$K_\nu(x, y) = \min\left\{1, \frac{|y|^{n-\alpha}}{|x|^{n-\alpha}}\right\} - \min\left\{1, \frac{|y|^{n-\alpha}}{|\nu|^{n-\alpha}}\right\}.$$

Then, since $K_\nu(x, y) = 0$ for $|y| > |\nu|$, we have

$$S_\alpha f(x) - S_\alpha f(\nu) = \int_{|y| \leq |\nu|} K_\nu(x, y) \frac{f(y)}{|y|^{n-\alpha}} dy. \tag{11}$$

If $|y| \le |\nu|$ then $K_\nu(x,y) \ge 0$, so

$$\frac{1}{\omega(B)}\int_B |S_\alpha f(x) - S_\alpha f(\nu)|dx \le \frac{1}{\omega(B)}\int_B\int_B K_\nu(x,y)\frac{|f(y)|}{|y|^{n-\alpha}}dydx$$
$$= \frac{1}{\omega(B)}\int_B\int_{|y|\le|x|} K_\nu(x,y)\frac{|f(y)|}{|y|^{n-\alpha}}dydx + \frac{1}{\omega(B)}\int_B\int_{|x|<|y|\le r} K_\nu(x,y)\frac{|f(y)|}{|y|^{n-\alpha}}dydx. \tag{12}$$

Now we estimate each term in (12).

If $|y| \le |x|$ then $K_\nu(x,y) \le |y|^{n-\alpha}|x|^{-(n-\alpha)}$. So, by (8) we have

$$\frac{1}{\omega(B)}\int_B\int_{|y|\le|x|} K_\nu(x,y)\frac{|f(y)|}{|y|^{n-\alpha}}dydx \le \frac{1}{\omega(B)}\int_B \frac{1}{|x|^{n-\alpha}}\int_B |f(y)|dydx$$
$$\le C\|f\|_{L^p(\omega^{-p})}|B|^{\delta}.$$

For the second term, since $0 \le K_\nu(x,y) \le 1$ and (8), we have

$$\frac{1}{\omega(B)}\int_B\int_{|x|<|y|\le r} K_\nu(x,y)\frac{|f(y)|}{|y|^{n-\alpha}}dydx \le \frac{1}{\omega(B)}\int_B \frac{1}{|x|^{n-\alpha}}\int_{|x|<|y|\le r} |f(y)|dydx$$
$$\le \frac{C}{\omega(B)}\int_B |f(y)||y|^{\alpha}dy \tag{13}$$
$$\le C\|f\|_{L^p(\omega^{-p})}|B|^{\delta}.$$

Then, by (12) and (13), we have proved

$$\frac{1}{\omega(B)|B|^{\delta}}\int_B |S_\alpha f(x) - S_\alpha f(\nu)|dx \le C\|f\|_{L^p(\omega^{-p})}, \tag{14}$$

for every ball B centered at the origin.

We now consider $B = B(x_0, r)$ with $r < |x_0|/8$. By (14) it is enough to consider only these balls B. Let $x \in B$ and $\nu = (|x_0| + r)x_0/|x_0|$. In the same way as (11), we have

$$S_\alpha f(x) - S_\alpha f(\nu) = \int_{|y|\le|\nu|} K_\nu(x,y)\frac{f(y)}{|y|^{n-\alpha}}dy.$$

Now, we note that if $|y| \le |\nu|$ then $K_\nu(x,y) \ge 0$. Applying the mean value theorem and using $|\nu| \sim |x|$, then

$$K_\nu(x,y) \le \frac{|y|^{n-\alpha}}{|x|^{n-\alpha}} - \frac{|y|^{n-\alpha}}{|\nu|^{n-\alpha}} \le C\frac{r|y|^{n-\alpha}}{|\nu|^{n-\alpha+1}}. \tag{15}$$

Thus, by (8) and $\omega \in D_\eta$, we have

$$\frac{1}{\omega(B)}\int_B |S_\alpha f(x) - S_\alpha f(\nu)|dx \le C\frac{r}{\omega(B)|\nu|^{n-\alpha+1}}\int_B\int_{|y|\le|\nu|} |f(y)|dydx$$
$$\le C\|f\|_{L^p(\omega^{-p})}\frac{r^{n+1}}{|\nu|^{n-\alpha+1+n/p}}\frac{\omega(B(0,|\nu|))}{\omega(B)} \tag{16}$$
$$\le C\|f\|_{L^p(\omega^{-p})}|B|^{\delta}.$$

Therefore, (14) and (16) complete the proof of the theorem.

Proof of Theorem 1.5: We begin showing the sufficient condition. Let $B = B(x_0, r)$ and let u, v, U and V as in (9) of the proof of Theorem 1.4. Then, we again have

$$\|S_\alpha f\|_{BMO^\delta(\omega)} \geq \frac{Cr^{n+1}}{\omega(B)|B|^\delta(|x_0|+r)^{n-\alpha+1}} \int_{B(0,|x_0|+r/2)} f(y)dy, \tag{17}$$

for every nonnegative function f in $BM_0^\gamma(\omega)$ such that $\mathrm{supp}(f) \subset B(0, |x_0|+r/2)$. Now, if $\gamma = 0$ we take $f(y) = \omega(y)\chi_{B(0,|x_0|+r/2)}(y)$ in (17) and since $\|f\|_{BM_0^\gamma(\omega)} \leq 1$, the boundedness of S_α and $\omega \in D_0$, we have $\omega \in D_\eta$.

If $\gamma > 0$, let $f(y) = P_{n\gamma}\left(\omega\chi_{B(0,|x_0|+r/2)}\right)(y)$, then $\|f\|_{BM_0^\gamma(\omega)} \leq C$ and

$$\begin{aligned} \int_{B(0,|x_0|+r/2)} f(y)dy &= C\int_{B(0,|x_0|+r/2)} \omega(t)((|x_0|+r/2)^{n\gamma} - |t|^{n\gamma})dt \\ &\geq C(|x_0|+r)^{n\gamma}\omega(B(0,(|x_0|+r/2)/2)). \end{aligned} \tag{18}$$

Therefore, using this function f in (17), the boundedness of S_α, (18) and $\omega \in D_0$, we have $\omega \in D_\eta$.

Now, let us show the necessary condition. Let $f \in BM_0^\gamma(\omega)$ such that $S_\alpha f(x)$ is finite for some $x \neq 0$ and let $\omega \in D_\eta$. Thus $S_\alpha f \in L^1_{loc}(\mathbb{R}^n)$ by (*ii*) of Lemma 4.7. We begin considering $B = B(0, r)$, $x \in B$ and $x \neq 0$. Let ν be such that $|\nu| = r$. In the same way as we did in (12), we have

$$\begin{aligned} \frac{1}{\omega(B)}\int_B |S_\alpha f(x) - S_\alpha f(\nu)|dx &\leq \frac{1}{\omega(B)}\int_B\int_{|y|\leq|x|} K_\nu(x,y)\frac{|f(y)|}{|y|^{n-\alpha}}dydx \\ &+ \frac{1}{\omega(B)}\int_B\int_{|x|<|y|\leq r} K_\nu(x,y)\frac{|f(y)|}{|y|^{n-\alpha}}dydx, \end{aligned} \tag{19}$$

where $K_\nu(x,y) = \min\left\{1, \frac{|y|^{n-\alpha}}{|x|^{n-\alpha}}\right\} - \min\left\{1, \frac{|y|^{n-\alpha}}{|\nu|^{n-\alpha}}\right\}$.

We estimate the first term of (19). Let $B_j = B(0, 2^{-j}r), j = 0, 1, \ldots$. Thus, since $K_\nu(x,y) \leq |y|^{n-\alpha}|x|^{-(n-\alpha)}$ for $|y| \leq |x|$ and (5), we have

$$\begin{aligned} \frac{1}{\omega(B)}\int_B\int_{|y|\leq|x|} K_\nu(x,y)\frac{|f(y)|}{|y|^{n-\alpha}}dydx &\leq \frac{1}{\omega(B)}\int_B \frac{1}{|x|^{n-\alpha}}\int_{|y|\leq|x|}|f(y)|dydx \\ &\leq C\|f\|_{BM_0^\gamma(\omega)}\frac{1}{\omega(B)}\int_B \frac{\omega(B(0,|x|))}{|x|^{n-\alpha-n\gamma}}dx \\ &\leq C\|f\|_{BM_0^\gamma(\omega)}\frac{r^{n\gamma+\alpha}}{\omega(B)}\sum_{j=0}^{\infty}\frac{\omega(B_j\backslash B_{j+1})}{2^{j(n\gamma+\alpha)}} \\ &\leq C\|f\|_{BM_0^\gamma(\omega)}\frac{|B|^\delta}{\omega(B)}\sum_{j=0}^{\infty}\omega(B_j\backslash B_{j+1}) \\ &= C\|f\|_{BM_0^\gamma(\omega)}|B|^\delta. \end{aligned} \tag{20}$$

For the second term of (19), since $0 \le K_\nu(x,y) \le 1$, we have

$$\begin{aligned}\frac{1}{\omega(B)}\int_B\int_{|x|\le|y|\le r} K_\nu(x,y)\frac{|f(y)|}{|y|^{n-\alpha}}dydx &\le \frac{1}{\omega(B)}\int_B\frac{|f(y)|}{|y|^{n-\alpha}}\int_{|x|\le|y|}1dxdy\\ &\le C\|f\|_{BM_0^\gamma(\omega)}|B|^\delta.\end{aligned} \tag{21}$$

Therefore, by (19)-(21) we have proved

$$\frac{1}{\omega(B)|B|^\delta}\int_B|S_\alpha f(x)-S_\alpha f(\nu)|dx \le C\|f\|_{BM_0^\gamma(\omega)}, \tag{22}$$

for every ball B centered at the origin.

We now consider $B = B(x_0, r)$ with $r < |x_0|/8$. By (22) it is enough to consider only these balls B. Let $x \in B$ and $\nu = (|x_0|+r)x_0/|x_0|$. In the same way as we obtained (11) and (15) in the previous proof, we have

$$S_\alpha f(x) - S_\alpha f(\nu) = \int_{|y|\le|\nu|} K_\nu(x,y)\frac{f(y)}{|y|^{n-\alpha}}dy$$

and $K_\nu(x,y) \le Cr|y|^{n-\alpha}|\nu|^{-(n-\alpha+1)}$. By $\omega \in D_\eta$, we have

$$\begin{aligned}\frac{1}{\omega(B)}\int_B|S_\alpha f(x)-S_\alpha f(\nu)|dx &\le C\frac{r^{n+1}}{\omega(B)|\nu|^{n-\alpha+1}}\int_{|y|\le|\nu|}|f(y)|dy\\ &\le C\|f\|_{BM_0^\gamma(\omega)}\frac{r^{n+1}}{|\nu|^{n-\alpha+1-n\gamma}}\frac{\omega(B(0,|\nu|))}{\omega(B)}\\ &\le C\|f\|_{BM_0^\gamma(\omega)}|B|^\delta.\end{aligned} \tag{23}$$

Therefore, (22) and (23), complete the proof of the theorem.

Let $x, \nu \in \mathbb{R}^n$, $\nu \ne 0$, then

$$\begin{aligned}|H_\alpha f(x) - H_\alpha f(\nu)| \le &\int_{|y|\le|\nu|}|f(y)|\left|\frac{1}{(|x|+|y|)^{n-\alpha}} - \frac{1}{(|\nu|+|y|)^{n-\alpha}}\right|dy\\ &+\int_{|y|>|\nu|}|f(y)|\left|\frac{1}{(|x|+|y|)^{n-\alpha}} - \frac{1}{(|\nu|+|y|)^{n-\alpha}}\right|dy.\end{aligned} \tag{24}$$

Proof of Theorem 1.6: We begin showing the sufficient condition. Let $B = B(x_0, r)$ and let u, v, U and V as in (9) of the proof of Theorem 1.4. Note that if $x \in U$, $z \in V$ then for all $y \in \mathbb{R}^n$,

$$\frac{1}{(|x|+|y|)^{n-\alpha}} - \frac{1}{(|z|+|y|)^{n-\alpha}} \ge C\frac{r}{(|x_0|+r+|y|)^{n-\alpha+1}}. \tag{25}$$

Hence, if f is a nonnegative function in $L^p(\omega^{-p})$ such that $\operatorname{supp}(f) \subset A(0,r,m)$ and taking $x_0 = 0$ in (25), we have

$$\|H_\alpha f\|_{BMO^\delta(\omega)} \ge \frac{Cr^{n+1}}{\omega(B)|B|^\delta}\int_{A(0,r,m)}\frac{f(y)}{|y|^{n-\alpha+1}}dy, \tag{26}$$

for every ball B centered at the origin.

Thus, taking $f_{m,j}(y) = |y|^{-(n-\alpha+1)/(p-1)}\omega^{p'}(y)\chi_{A_{m,j}}(y)$ in (26) where $A_{m,j} = A(0,r,m)\cap\{y : 1/j\le\omega(y)<j\}, m,j = 1,2, \ldots$, using the boundedness of H_α and letting $m\to\infty, j\to\infty$ we obtain that $\omega\in \mathrm{H}_0(\alpha,p)$.

On the other hand, if f is a nonnegative function in $L^p(\omega^{-p})$ such that $\mathrm{supp}(f)\subset B(0,2(|x_0|+r))$, then by (25)

$$\|H_\alpha f\|_{BMO^\delta(\omega)} \ge \frac{Cr^{n+1}}{\omega(B)|B|^\delta(|x_0|+r)^{n-\alpha+1}}\int_{B(0,2(|x_0|+r))} f(y)dy. \tag{27}$$

Thus, taking $f_j(y) = \omega^{p'}(y)\chi_{A_j}(y)$ in (27) where $A_j = B(0,2(|x_0|+r))\cap\{y : 1/j\le\omega(y)<j\}, j = 1,2, \ldots$, and using the boundedness of H_α, we have

$$\left(\int_{A_j}\omega^{p'}(y)dy\right)^{1/p'}\le C\left(\frac{|x_0|+r}{r}\right)^{n-\alpha+1}\frac{\omega(B)}{|B|^{1/p}}.$$

Letting $j\to\infty$ and taking $x_0 = 0$ in the last expression, we can obtain that $\omega\in RH_0(p')$. Then, applying Hölder's inequality, we obtain $\omega\in D_\eta$.

Now, let us show the necessary condition. Let $f\in L^p(\omega^{-p})$ such that $H_\alpha f(x)$ is finite for some $x\ne 0$ and let ω such that $\omega\in \mathrm{H}_0(\alpha,p)\cap RH_0(p')\cap D_\eta$. Hence $H_\alpha f\in L^1_{loc}(\mathbb{R}^n)$ since (i) of Lemma 4.7. We begin considering $B = B(0,r)$, $x\in B$ and $x\ne 0$. Let ν be such that $|\nu| = r$. We estimate the two terms of (24). By (8), we have

$$\begin{aligned}\frac{1}{\omega(B)}\int_B\int_{|y|\le|\nu|}&\left|\frac{f(y)}{(|x|+|y|)^{n-\alpha}}-\frac{f(y)}{(|\nu|+|y|)^{n-\alpha}}\right|dydx\\ &\le\frac{C}{\omega(B)}\int_B\int_B\frac{|f(y)|}{|x|^{n-\alpha}}dydx\\ &\le C\|f\|_{L^p(\omega^{-p})}\int_B\frac{1}{|x|^{n-\alpha+n/p}}dx = C\|f\|_{L^p(\omega^{-p})}|B|^\delta.\end{aligned} \tag{28}$$

To analyze the second term of (24), we use the mean value theorem, then

$$\left|\frac{1}{(|x|+|y|)^{n-\alpha}}-\frac{1}{(|\nu|+|y|)^{n-\alpha}}\right|\le C\frac{r}{|y|^{n-\alpha+1}}.$$

Thus, by (i) of Lemma 4.6

$$\begin{aligned}\frac{1}{\omega(B)}\int_B\int_{|y|>|\nu|}\left|\frac{f(y)}{(|x|+|y|)^{n-\alpha}}-\frac{f(y)}{(|\nu|+|y|)^{n-\alpha}}\right|dydx &\le\frac{Cr}{\omega(B)}\int_B\int_{B^c}\frac{|f(y)|}{|y|^{n-\alpha+1}}dydx\\ &\le C\|f\|_{L^p(\omega^{-p})}|B|^\delta.\end{aligned} \tag{29}$$

Therefore, by (24)–(29), we have proved

$$\frac{1}{\omega(B)|B|^\delta}\int_B|H_\alpha f(x)-H_\alpha f(\nu)|dx\le C\|f\|_{L^p(\omega^{-p})}, \tag{30}$$

for every ball B centered at the origin.

We now consider $B = B(x_0, r)$ with $r < |x_0|/8$. By (28) it is enough to consider only these balls B. Let $x \in B$ and $\nu = (|x_0|+r)x_0/|x_0|$, then $|\nu| \sim |x|$ and $|x| \sim |x_0|$. Using $|y| \le |\nu|$ and the mean value theorem

$$\left|\frac{1}{(|x|+|y|)^{n-\alpha}} - \frac{1}{(|\nu|+|y|)^{n-\alpha}}\right| \le C\frac{r}{|x_0|^{n-\alpha+1}}.$$

Then, by (8) and $\omega \in D_\eta$

$$\begin{aligned}\frac{1}{\omega(B)}\int_B\int_{|y|\le|\nu|} |f(y)|\left|\frac{1}{(|x|+|y|)^{n-\alpha}} - \frac{1}{(|\nu|+|y|)^{n-\alpha}}\right| dydx \\ \le \frac{Cr^{n+1}}{\omega(B)|x_0|^{n-\alpha+1}}\int_{|y|\le|\nu|}|f(y)|dy \\ \le C\|f\|_{L^p(\omega^{-p})}\frac{r^{n+1}}{|x_0|^{n-\alpha+1+n/p}}\omega(B(0,|\nu|)) \\ = C\|f\|_{L^p(\omega^{-p})}|B|^{\delta}.\end{aligned} \tag{31}$$

Now, using the mean value theorem

$$\left|\frac{1}{(|x|+|y|)^{n-\alpha}} - \frac{1}{(|\nu|+|y|)^{n-\alpha}}\right| \le C\frac{r}{|y|^{n-\alpha+1}}.$$

Then, by (i) of Lemma 4.6

$$\begin{aligned}\frac{1}{\omega(B)}\int_B\int_{|y|>|\nu|}\left|\frac{f(y)}{(|x|+|y|)^{n-\alpha}} - \frac{f(y)}{(|\nu|+|y|)^{n-\alpha}}\right| dydx \le \frac{Cr}{\omega(B)}\int_B\int_{B^c}\frac{|f(y)|}{|y|^{n-\alpha+1}}dydx \\ = C\|f\|_{L^p(\omega^{-p})}|B|^{\delta}.\end{aligned} \tag{32}$$

Therefore, by (24) with $\nu = (|x_0|+r)x_0/|x_0|$, (31) and (32), we have

$$\frac{1}{\omega(B)|B|^{\delta}}\int_B |H_\alpha f(x) - H_\alpha f(\nu)|dx \le C\|f\|_{L^p(\omega^{-p})},$$

for every ball $B = B(x_0, r)$ considered. This completes the proof of the theorem.

Proof of Theorem 1.7: We begin showing the sufficient condition. Let $B = B(x_0, r)$ and let u, v, U and V as in (9) of the proof of Theorem 1.4. Then, as in (26) of the proof of Theorem 4 (with $x_0 = 0$), we again have

$$\|H_\alpha f\|_{BMO^{\delta}(\omega)} \ge \frac{Cr^{n+1}}{\omega(B)|B|^{\delta}}\int_{A(0,r,m)}\frac{f(y)}{|y|^{n-\alpha+1}}dy \tag{33}$$

for every nonnegative function f in $BM_0^{\gamma}(\omega)$ such that $\operatorname{supp}(f) \subset A(0,r,m)$ and for every ball B centered at the origin.

Thus, taking $f(y) = |y|^{n\gamma}\omega(y)\chi_{A(0,r,m)}(y)$ in (33), using that $\|f\|_{BM_0^{\gamma}(\omega)} \le 1$, the boundedness of H_α and letting $m \to \infty$, we have that $\omega \in \mathrm{H}_0(\alpha + n\gamma, \infty)$.

On the other hand, as in (27) of the proof of Theorem 1.6 we again have

$$\|H_{\alpha}f\|_{BMO^{\delta}(\omega)} \geq \frac{Cr^{n+1}}{\omega(B)|B|^{\delta}(|x_0|+r)^{n-\alpha+1}} \int_{B(0,2(|x_0|+r))} f(y)dy, \tag{34}$$

for every nonnegative function f in $BM_0^{\gamma}(\omega)$ such that $\operatorname{supp}(f) \subset B(0,2(|x_0|+r))$ and for every ball $B = B(x_0, r)$.

If $\gamma = 0$, we take $f(y) = \omega(y)\chi_{B(0,2(|x_0|+r))}(y)$ in (4.34) and since $\|f\|_{BM_0^{\gamma}(\omega)} \leq 1$ and the boundedness of H_{α}, we have $\omega \in D_{\eta}$.

If $\gamma > 0$, let $f(y) = P_{n\gamma}\left(\omega\chi_{B(0,2(|x_0|+r))}\right)(y)$ then $\|f\|_{BM_0^{\gamma}(\omega)} \leq C$ and as in (4.18) of the proof of Theorem 1.5, we have

$$\int_{B(0,2(|x_0|+r))} f(y)dy \geq C(|x_0|+r)^{n\gamma}\omega(B(0,|x_0|+r)).$$

Therefore, using this function f in (34) and the boundedness of H_{α}, we have $\omega \in D_{\eta}$.

Now, let us show the necessary condition. Let $f \in BM_0^{\gamma}(\omega)$ such that $H_{\alpha}f(x)$ is finite for some $x \neq 0$ and let $\omega \in \mathrm{H}_0(\alpha + n\gamma, \infty) \cap D_{\eta}$. Hence $H_{\alpha}f \in L^1_{loc}(\mathbb{R}^n)$ by (ii) of Lemma 4.7. We begin considering $B = B(0,r)$, $x \in B$ and $x \neq 0$. Let ν be such that $|\nu| = r$. We estimate the two terms of (24). Then,

$$\begin{aligned}
&\frac{1}{\omega(B)}\int_B\int_{|y|\leq|\nu|}|f(y)|\left|\frac{1}{(|x|+|y|)^{n-\alpha}} - \frac{1}{(|\nu|+|y|)^{n-\alpha}}\right|dydx \\
&\leq \frac{C}{\omega(B)}\int_B \frac{1}{|x|^{n-\alpha}}\int_{|y|\leq|\nu|} f(y)dydx \\
&\leq C\|f\|_{BM_0^{\gamma}(\omega)}\frac{1}{\omega(B)}\int_B \frac{\omega(B(0,|\nu|))|\nu|^{n\gamma}}{|x|^{n-\alpha}}dx \\
&\leq C\|f\|_{BM_0^{\gamma}(\omega)}|B|^{\delta}.
\end{aligned} \tag{35}$$

For the second term of (24), using the mean value theorem

$$\left|\frac{1}{(|x|+|y|)^{n-\alpha}} - \frac{1}{(|\nu|+|y|)^{n-\alpha}}\right| \leq C\frac{r}{|y|^{n-\alpha+1}}. \tag{36}$$

Then, by (ii) of Lemma 4.6

$$\begin{aligned}
&\frac{1}{\omega(B)}\int_B\int_{|y|>|\nu|}|f(y)|\left|\frac{1}{(|x|+|y|)^{n-\alpha}} - \frac{1}{(|\nu|+|y|)^{n-\alpha}}\right|dydx \\
&\leq \frac{Cr}{\omega(B)}\int_B\int_{B^c}\frac{|f(y)|}{|y|^{n-\alpha+1}}dydx \\
&\leq C\|f\|_{BM_0^{\gamma}(\omega)}|B|^{\delta}.
\end{aligned} \tag{37}$$

Therefore, by (24) and (35)–(37), we have proved

$$\frac{1}{\omega(B)|B|^{\delta}}\int_B |H_{\alpha}f(x) - H_{\alpha}f(\nu)|dx \leq C\|f\|_{BM_0^{\gamma}(\omega)}, \tag{38}$$

for every ball B centered at the origin.

We now consider $B = B(x_0, r)$ with $r < |x_0|/8$. By (33) it is enough to consider only these balls B. Let $x \in B$ and $\nu = (|x_0| + r)x_0/|x_0|$, then $|\nu| \sim |x|$ and $|x| \sim |x_0|$. If $|y| \leq |\nu|$, by the mean value theorem

$$\left|\frac{1}{(|x| + |y|)^{n-\alpha}} - \frac{1}{(|\nu| + |y|)^{n-\alpha}}\right| \leq C\frac{r}{|x_0|^{n-\alpha+1}}.$$

Then, since $\omega \in D_\eta$, we have

$$\begin{aligned}
&\frac{1}{\omega(B)}\int_B\int_{|y| \leq |\nu|} |f(y)|\left|\frac{1}{(|x| + |y|)^{n-\alpha}} - \frac{1}{(|\nu| + |y|)^{n-\alpha}}\right| dydx \\
&\leq \frac{Cr^{n+1}}{\omega(B)|x_0|^{n-\alpha+1}}\int_{|y| \leq |\nu|} |f(y)|dy \\
&\leq \|f\|_{BM_0^\gamma(\omega)}|B|^\delta.
\end{aligned} \tag{39}$$

On the other hand, using again the mean value theorem as in (36) and (*ii*) of Lemma 4.6, we get

$$\begin{aligned}
&\frac{1}{\omega(B)}\int_B\int_{|y| > |\nu|} |f(y)|\left|\frac{1}{(|x| + |y|)^{n-\alpha}} - \frac{1}{(|\nu| + |y|)^{n-\alpha}}\right| dydx \\
&\leq \frac{Cr^{n+1}}{\omega(B)}\int_{|y| > |\nu|} \frac{|f(y)|}{|y|^{n-\alpha+1}}dy \\
&\leq C\|f\|_{BM_0^\gamma(\omega)}\frac{r^{n+1}}{\omega(B)}\frac{\omega(B(0, |\nu|))}{|\nu|^{n\eta}} \\
&\leq C\|f\|_{BM_0^\gamma(\omega)}|B|^\delta.
\end{aligned} \tag{40}$$

Thus, by (24) and (39)–(40), we have proved

$$\frac{1}{\omega(B)|B|^\delta}\int_B |H_\alpha f(x) - H_\alpha f(\nu)|dx \leq C\|f\|_{BM_0^\gamma(\omega)},$$

for every ball $B = B(x_0, r)$ considered. This completes the proof of the theorem.

5. Conclusions

As a conclusion to this chapter, we have given necessary and sufficient conditions for the generalized Calderón and Hilbert operators to be bounded from weighted Lesbesgue spaces into suitable weighted *BMO* and Lipschitz spaces. Then, we have obtained results on the boundedness of these operators from L^∞ into *BMO*, even in the unweighted case for the Hilbert operator. The class of weights involved are close to the doubling and reverse Hölder conditions related to the Muckenhoupt's classes.

The study of the weighted boundedness for integral operators on function spaces, like the one we develop in this chapter, is one of the main research fields in harmonic analysis. In particular, it has had a profound influence in partial differential equations, several complex variables, and number theory. Evidence of such success and importance is the pioneering work of leading mathematicians Bourgain, Zygmund, Calderón, Muckenhoupt, Wheeden, C. Fefferman, Stein, Ricci, Tao and so on.

Acknowledgements

I would like to deeply thank Dr. Hammad Khalil for considering me to write this chapter. Also, I would like to thank Professors E. Ferreyra and B. Viviani who published the article [7] and for many helpful discussions.

Author details

Guillermo J. Flores
Centro de Investigación y Estudios de Matemática (CIEM), Facultad de Matemática, Astronomía, Física y Computación (FaMAF), Universidad Nacional de Córdoba (UNC), Córdoba, Argentina

*Address all correspondence to: guilleflores@unc.edu.ar

References

[1] Hardy GH. Notes on some points in the integral calculus. Messenger of Mathematics. 1918;**48**:107-112

[2] Hardy GH. Note on a theorem of Hilbert. Mathematische Zeitschrift. 1920;**6**:314-317. DOI: 10.1007/BF01199965

[3] Hardy GH, Littlewood JE. Notes on the theory of series (VI): Two inequalities. Journal of the London Mathematical Society. 1927;**2**(3):196-201. DOI: 10.1112/jlms/s1-2.3.196

[4] Hardy GH, Littlewood JE, Pólya G. Inequalities. Cambridge: Cambridge Mathematical Library, Cambridge University Press; 1988

[5] Bastero J, Milman M, Ruiz FJ. On the connection between weighted norm inequalities, commutators and real interpolation. Memoirs of the American Mathematical Society. 2001;**154**(731): viii. DOI: 10.1090/memo/0731

[6] Muckenhoupt B. Hardy's inequality with weights. Studia Mathematica. 1972; **44**:31-38

[7] Duoandikoetxea J, Martín-Reyes FJ, Ombrosi S. Calderón weights as Muckenhoupt weights. Indiana University Mathematics Journal. 2013; **62**(3):891-910. DOI: 10.1512/iumj.2013.62.4971

[8] Harboure E, Salinas O, Viviani B. Boundedness of the fractional integral on weighted Lebesgue and Lipschitz spaces. Transactions of the American Mathematical Society. 1997;**349**(1): 235-255. DOI: 10.1090/S0002-9947-97-01644-9

[9] Muckenhoupt B, Wheeden R. Weighted norm inequalities for fractional integrals. Transactions of the American Mathematical Society. 1974;**192**:261-274. DOI: 10.2307/1996833

[10] Ferreyra E, Flores G, Viviani B. Weighted Lebesgue and BMO norm inequalities for the Calderón and Hilbert operators. Mathematische Zeitschrift. 2020;**294**:503-518. DOI: 10.1007/s00209-019-02298-6

[11] Dieudonné J. Treatise on analysis. In: Pure and Applied Mathematics. Boston: Academic Press; 1993

[12] Duoandikoetxea J. Fractional integrals on radial functions with applications to weighted inequalities. Annali di Matematica Pura ed Applicata. 2013;**4**:553-568. DOI: 10.1007/s10231-011-0237-7

[13] Ferreyra E, Flores G. Weighted estimates for integral operators on local BMO type spaces. Mathematische Nachrichten. 2015;**288**(8–9):905-916. DOI: 10.1002/mana.201400121

[14] Harboure E, Chicco Ruiz A. BMO spaces related to Laguerre semigroups. Mathematische Nachrichten. 2014;**287**(2–3):254-280. DOI: 10.1002/mana.201200227

[15] Harboure E, Segovia C, Torrea JL, Viviani B. Power weighted L-inequalities for Laguerre–Riesz transforms. Arkiv för Matematik. 2008; **46**(2):285-313. DOI: 10.1007/s11512-007-0052-y

[16] Nowak A, Stempak K. Weighted estimates for the Hankel transform transplantation operator. Tohoku Mathematical Journal. 2006;**2**:277-301. DOI: 10.1007/s00009-021-01951-x

[17] Weyl H. Singuläre Integralgleichungen. Mathematische Annalen. 1908;**66**(3):273-324. DOI: 10.1007/BF01450690

[18] Muckenhoupt B, Wheeden RL. Weighted bounded mean oscillation and the Hilbert transform. Studia Mathematica. 1975;**54**(3):221-237. Available from: http://eudml.org/doc/217982

[19] Grafakos L. Modern Fourier Analysis. 2nd Ed. New York: Springer-Verlag; 2009. DOI: 10.1007/978-0-387-09434-2

[20] Fefferman C, Muckenhoupt B. Two nonequivalent conditions for weight functions. Proceedings of American Mathematical Society. 1974;**45**:99-104. Available from: www.jstor.org/stable/2040615

Section 2

Applications

Chapter 4

Effect of Titanium Oxide Nanofluid over Cattaneo-Christov Model

Hammad Khalil, Tehseen Zahra, Zaffer Elahi and Azeem Shahzad

Abstract

The proposed chapter deals with the study of heat transfer development of titanium oxide nanofluid of platelet shape nanoparticles over a vertical stretching cylinder. The set of nonlinear equations is obtained using suitable transformation on the governing equations that are then solved with numerical scheme BVP4C. The obtained results are interpreted graphically and numerically. The effects of Prandtl, Eckert, and unsteadiness parameters on temperature distribution are depicted. Moreover the skin friction and Nusselt number are also computed.

Keywords: *Cattaneo*-Christov model, heat transfer, vertical cylinder

1. Introduction

Cattaneo-Christov model is an improved version of Fourier law as Fourier law does not detect the initial temperature disturbance; to overcome this ambiguity, Cattaneo added a thermal relaxation parameter. This parameter covers the ambiguity of Fourier law. The classical Fourier law is obtained while vanishing the relaxation parameter [1]. Cattaneo-Christov heat flux model gives us heat transfer rate in stretching cylinders as well as sheets. Heat transfer is a wonderful natural phenomenon that occurs when two bodies have a thermal difference until both bodies are at thermal equilibrium. The Cattaneo-Christov model is in the form of a heat equation. The thermal convection effect is studied using the Christov heat model in conjunction with the Cattaneo heat model [2]. It has been realized that the development of stretchy surfaces and the flow field that surrounds them speaks to a variety of technological and industrial applications, such as paper making, glass blowing, crystal growth, and aerodynamic plastic sheet extrusion [3]. Heat transfer is a common natural occurrence as long as there is a temperature differential between things or between various regions of the same object, heat transfer will occur. As a result, a lot of effort has gone into predicting the heat transport behavior. In several starting and boundary problems, the uniqueness and structural stability of the solutions for the temperature governing equations using the Cattaneo-Christov heat flow model have been demonstrated. The chapter released uses the Cattaneo-Christov heat flux model to analyze the flow and heat transfer of upper-convective Maxwell fluid across a stretching sheet [4]. Efforts have been undertaken to increase the thermal efficiency of processes during the last many decades. On the one hand, there has been an attempt to lower the size of the equipment by increasing the thermal exchange surface, such as with fins, and on the other hand, novel fluid exchangers with higher thermal conductivity have been developed. Different NPS types (metallic, nonmetallic, and carbon based) have been synthesized and dispersed in conventional fluids such as water, oil, or ethylene glycol referred to as nanofluids since the advent of nanotechnology and the possibility of synthesizing materials on a nanometric scale [5]. The boundary layer flow and heat transfer caused by stretching flat plates or cylinders are both practical and theoretically interesting in fiber technology and extrusion operations. This method is used to produce polymer sheets and plastic films. The cooling of an infinite metallic plate in a cooling bath, the boundary layer along material handling conveyors, the aerodynamic extrusion of plastic sheets, the boundary layer along a liquid film in condensation processes, paper production, glass blowing, metal spinning and drawing plastic films, and polymer extrusion are all examples of boundary layers [6]. The aim of this chapter is to manipulate the heat transfer rate of titanium oxide nanofluid with the Cattaneo-Christov heat flux model over a vertical stretching cylinder.

2. Mathematical formulation

In the coordinate plane, assume that the cylinder is taken in the vertical direction along the z-axis, and the r-axis is normal to the axis of the cylinder. Consider the fluid is moving with surface velocity,

$$U_w = \frac{bz}{1-\alpha t}$$

In the direction of stretching cylinder under the external magnetic field defined by

$$B(t) = \frac{B_0}{\sqrt{1-\alpha t}}$$

Further, suppose $u = u(r, z, t)$ and $w = w(r, z, t)$ be the velocity components along the respective axes of the coordinate plane, while $T = T(r, z, t)$ denotes the temperature of the nanofluid, as shown in **Figure 1**.

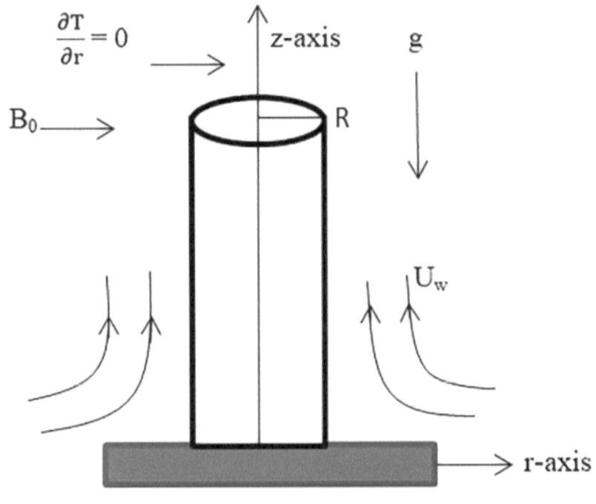

Figure 1.
Representation of the dilemma.

Let the base fluid (water) and platelet-shaped NPS in thermal equilibrium. Under these assumptions, the equation of continuity, momentum equation, and energy equations are obtained, which are as follows:

$$\frac{\partial(ru)}{\partial r}+\frac{\partial(rw)}{\partial z}=0 \tag{1}$$

$$+\frac{u\,\partial w}{\partial r}+\frac{w\,\partial w}{\partial z}=\frac{\nu}{r}\frac{\partial}{\partial r}\left(\frac{r\,\partial w}{\partial r}\right)-\frac{\sigma}{\rho}\,B_o^{\,2}\,w+g\beta(T-T\infty) \tag{2}$$

$$\frac{\partial T}{\partial t}+\frac{u\,\partial T}{\partial r}+\frac{w\,\partial T}{\partial z}=\frac{\nu}{Cp}\left(\frac{\partial w}{\partial r}\right)^2+\lambda_1\left[\frac{2\nu}{Cp}\left(\frac{\partial w}{\partial r}\right)\left(\frac{\partial^2 w}{\partial t\partial r}+u\frac{\partial^2 w}{\partial r^2}\right)-\left(\frac{\partial^2 T}{\partial t^2}\right.\right.$$
$$+u^2\frac{\partial^2 T}{\partial r^2}+w^2\frac{\partial^2 T}{\partial z^2}2u\frac{\partial^2 T}{\partial t\partial r}+2w\frac{\partial^2 T}{\partial t\partial z}+2uw\frac{\partial^2 T}{\partial z\partial r}$$
$$\left.\left.+\left(\frac{\partial u}{\partial t}+u\frac{\partial u}{\partial r}+w\frac{\partial u}{\partial z}\right)\frac{\partial T}{\partial r}+\left(\frac{\partial w}{\partial t}+u\frac{\partial w}{\partial r}+w\frac{\partial w}{\partial z}\right)\frac{\partial T}{\partial z}\right)\right]. \tag{3}$$

Subject to the boundary conditions

$$\mathrm{u}=0,\mathrm{w}=U_{\mathrm{w}},\frac{\partial T}{\partial r}=0 \text{ at } \mathrm{r}=\mathrm{R},$$
$$w\to o, T\to T_\infty \text{ as } r\to\infty \tag{4}$$

The thermophysical properties of density (ρnf), dynamic viscosity (μnf), electric conductivity (σnf), diffusivity (αnf), and heat capacity (ρCp) can be defined in Refs. [7, 8], while the ratio of thermal conductivity of nanofluid and base fluid is given by the following equation:

$$\frac{k_{nf}}{k_f}=\left[\frac{k_s+(m-1)k_f+(m-1)(k_s-k_f)\phi}{k_s+(m-1)k_f-(k_s-k_f)\phi}\right] \tag{5}$$

where ϕ denotes volume-fraction of NPS.

The thermophysical properties of titanium nanofluid with base fluid as water [9] are given in **Table 1**, while the viscosity coefficients $A1$, $A2$, and shape factor m values of TiO_2 nanofluid [10] are listed in **Table 2**.

Base	Density (kg/m^3)	Thermal conductivity (W/m K)	Specific heat (J/kg K)
TiO_2	3900	8.4	0.8692
H_2O	997.1	0.613	4179

Table 1.
Thermophysical properties of base fluid and TiO_2 nanoparticles.

Platelet		
A_1	A_2	m
37.1	612.6	5.72

Table 2.
Viscosity and shape factor values of platelet-shaped nanoparticles.

Introducing the transformations, as

$$T = T_\infty + (Tw - T_\infty)\,\theta(\eta), \eta = \left(\sqrt{\frac{c}{\nu(1-\alpha t)}}\right)\left(\frac{r^2 - R^2}{2R}\right), \psi = \left(\sqrt{\frac{c\nu}{(1-\alpha t)}}\right) zrf(\eta)\partial\psi \tag{6}$$

where ψ is the stream function (describes the flow pattern) and is defined as $u = \frac{-1}{r}\frac{\partial\psi}{\partial r}$ and $w = \frac{1}{r}\frac{\partial\psi}{\partial r}$. The governing Eqs. (2)–(5) have been transformed to Eqs. (8)–(10) using similarity variables in Eq. (7), as

$$\begin{aligned} &\epsilon_1(1+2C\eta)f'''(\eta) + 2\epsilon_1 Cf''(\eta) - \epsilon_3 M\, f(\eta) \\ &\quad + \left[f(\eta)f''^{(\eta)} - f'^2(\eta) - S\left(f'(\eta) + \frac{\eta}{2}f''(\eta)\right)\right] + \lambda\theta(\eta) = 0, \end{aligned} \tag{7}$$

$$\begin{aligned} &(1+2C\eta)\left(\epsilon_1 Ecf''^2(\eta) + \frac{\epsilon_2}{Pr}\theta''(\eta)\right) + \frac{2\epsilon_2}{Pr}C\theta'^{(\eta)} + f(\eta)\theta'^{(\eta)} - S\left(2\theta(\eta) + \frac{\eta}{2}\,f''(\eta)\right) \\ &+\beta\left[\begin{array}{l} \epsilon_1 Ec(1+2\eta)\left(3Sf''^2(\eta) + S\eta f''(\eta)f'''(\eta) - 2f(\eta)f''(\eta)f'''(\eta)\right)2\epsilon_1 EcCf(\eta)f''^2(\eta) \\ -\left(S^2\left(\begin{array}{l} 6\theta(\eta) + \frac{11}{4}\theta''(\eta)f(\eta) + S\left(\begin{array}{l} 5f'(\eta)\theta(\eta) - \frac{11}{2}f(\eta)\theta'(\eta) - \eta f(\eta)\theta''(\eta) \\ +\left(\eta - \frac{1}{2}\right)f'(\eta)\theta''(\eta) + \frac{\eta}{2}f''(\eta)\theta(\eta) \end{array}\right) \\ +f'(\eta)\theta''(\eta) - f(\eta)f'(\eta)\theta'(\eta) - f(\eta)f''(\eta)\theta(\eta) + f'^2(\eta)\theta(\eta) \end{array}\right)\right) \end{array}\right] \\ &-f'^{(\eta)}\theta(\eta). \end{aligned} \tag{8}$$

where β is the thermal relaxation parameter and is given by,

$$\beta 1 = \frac{c\lambda 1}{(1-\alpha t)} \tag{9}$$

Under the boundary condition,

$$f(0) = 0, f'(0) = 1, \theta'(0) = \frac{-k_f}{k_{nf}}\,\gamma(1 - \theta(0)) \text{ at } \eta = R$$

$$f'(\eta) = 0, \theta(0) \text{ as } \eta \to \infty \tag{10}$$

Now the dimensionless constants, such as Ec, Pr, ϕ, M, and S, and that of $\epsilon 1$, $\epsilon 2$, and $\epsilon 3$ are used frequently in the above equations, defined in Ref. [11]. For various values of dimensionless parameters, the value of the local Nusselt number is shown in **Table 3**. Nusselt number can be defined as,

$$\mathrm{Nu} = \frac{zk_{nf}}{k_f(T_{w-}T_\infty)}\left[\frac{\partial T}{\partial r}\right]_{\mathrm{r=R}} \tag{11}$$

The non-dimensionless form of Eq. (11), using Eq. (6), as

$$Re^{\frac{-1}{2}}\,Nu = -\frac{k_{nf}}{k_f}\,\theta'(0) \tag{12}$$

3. Method of solution

To find the numerical solution of a nonlinear system (7) and (8), the set of first-order linear equations is obtained by considering the following assumptions. By putting these assumptions in the above equations, we get first-order linear equations, which are then used in MATLAB by using BVP4C scheme to get numerical and graphical results.

$$y_1 = f, \tag{13}$$

$$y_1' = y_2, \tag{14}$$

$$y_2' = y_3, \tag{15}$$

$$y_3' = g_1, \tag{16}$$

$$\theta = y_4, \tag{17}$$

$$y_4' = g_1, \tag{18}$$

$$y_5' = g_2, \tag{19}$$

$$y_1(0) = 0, y_2(0) = 1, y_4(0) - 1 = 0 \text{ at } \eta = 0 \tag{20}$$

$$y_2(\eta) = 0, y_4(\eta) = 0.\text{as } \eta \to \infty \tag{21}$$

	Physical parameters	**Platelet**
Ec	***Pr***	$\mathbf{Re^{\frac{-1}{2}}\,Nu}$
0.0	6.0	0.037666741
0.5	—	0.038924615
1.0	—	0.039829280
1.0	4.0	0.042765482
—	6.0	0.038924615
—	8.0	0.036302577

Table 3.
Nusselt number of platelet shape nanoparticle.

where

$$g_1 = \frac{1}{\epsilon_1(1+2\eta C)}\left(\epsilon_3 M y_2 - 2\epsilon_1 C y_3 - \lambda y_4 - \left(y_1 y_3 - y_2^2 - S\left(y_2 + \frac{\eta}{2} y_3\right)\right)\right), \tag{22}$$

and

$$g_2 = \frac{1}{\frac{(1+2\eta C)\epsilon_3}{Pr} - \beta_1\left(\frac{(S\eta)^2}{4} - S\eta y_1 + y_1^2\right)}\left(S\left(y_4 + \frac{\eta}{2} y_5\right) - y_1 + y_2 y_4 - \left(\frac{2\epsilon_2 C y_5}{Pr}\right)\right.$$
$$-(1+2\eta C)\epsilon_1 Ec y_3^2 - \beta_1\left(\epsilon_1 Ec(1+2\eta C)\left(3Sy_3^2\right) + Sty_3 y_3' - 2\epsilon_1 EcCy_1 y_3^2 - 2y_1 y_3 y_3'\right)$$
$$\left.-\left(S^2\left(6y_4 + \frac{11}{4}\eta y_5\right) - S\left(5y_2 y_4 - \frac{11}{2} y_1 y_5\right) + \left(\eta - \frac{1}{2} y_2 y_5 + \frac{\eta}{2} y_3 y_4\right) + y_1 y_3 y_5 + y_2^2 y_4\right)\right)\right). \tag{23}$$

g_1 and g_2 are the obtained first-order linear equations.

4. Analysis of results

Obtained numerical results and the effect of various parameters on temperature profile are obtained in both numerical and graphical form and discussed in detail.

4.1 Graphical analysis

Figure 2 describes the influence of Eckert number for platelet shape nanoparticle. By varying the value of Eckert, it can be seen that the temperature is increasing. Physically, it can be seen that the Eckert number enhances the thermal conductivity of the fluid. **Figure 3** shows that the increase in the value of the Prandtl number results

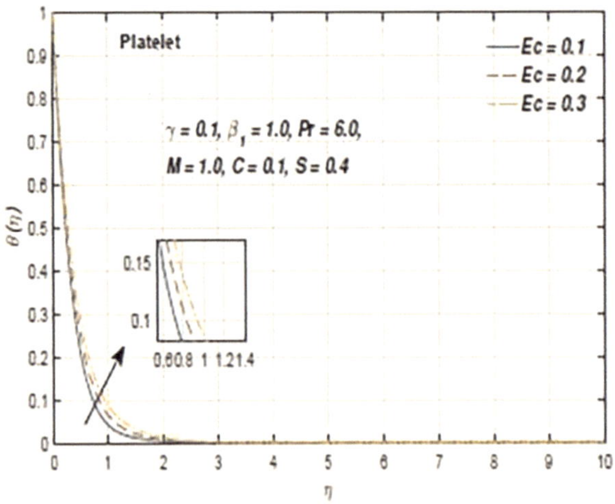

Figure 2.
Effect of Eckert number on the temperature profile.

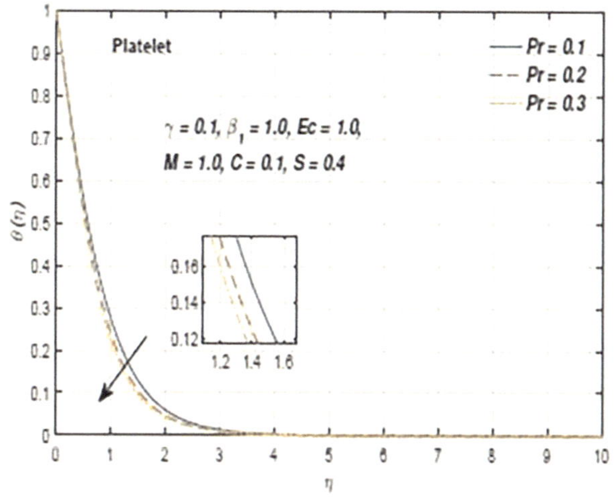

Figure 3.
Effect of Prandtl number on the temperature profile.

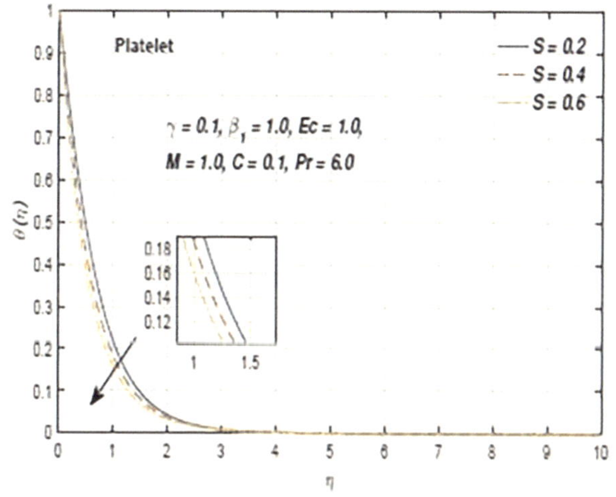

Figure 4.
Effect of unsteadiness parameter number on the temperature profile.

in deceleration in temperature because of the reduction in thermal diffusivity. As the unsteadiness parameter increases, the temperature gradually decreases, as shown in **Figure 4**. Physically the value of the unsteadiness parameter is grown up, and the thickness of the thermal boundary layer decreases, which results in a decline in the temperature profile.

4.2 Numerical results

The heat transfer rate is calculated for platelet shape nanoparticles, which is given in **Table 3**. It is inferred that with the rise in Eckert number, Nusselt number increases while the reverse trend is seen for Prandtl number

5. Conclusion

By numerical computation, the effect of platelet shape nanoparticle on TiO_2 nanofluid over a vertical stretching cylinder is seen in this chapter. Influence of different physical parameters, such as Eckerd and Prandtl numbers on temperature profile, is examined both graphically and numerically. The Nusselt number increases for Eckert number Ec, which decreases for Prandtl number Pr. Graphical result shows acceleration in temperature profile, while the reverse trend is found in **Figure 2**.

Nomenclature

u, w	velocity components along r, z directions (m/s)
αf, αnf	thermal diffusion of base fluid and nanofluid (m^2/s)
ρf, ρnf	density of base fluid and nanofluid (kg/m^3)
μf, μnf	viscosity of base fluid and nanofluid (kg m/s)
νf, νnf	kinematic viscosity of base fluid and nanofluid (m^2/s)
σnf	electrical conductivity (−)
Uw	surface velocity (m/s)
σnf	electrical conductivity (−)
$(\rho Cp)\ nf$	heat capacity of nanofluid (−)

Author details

Hammad Khalil[1]*, Tehseen Zahra[2], Zaffer Elahi[2] and Azeem Shahzad[2]

1 University of Education, Pakistan

2 University of Engineering and Technology, Taxila, Pakistan

*Address all correspondence to: hammad.khalil@ue.edu.pk

References

[1] Mustafa M. Cattaneo-Christov heat flux model for rotating flow and heat transfer of upper-convected Maxwell fluid. AIP Publishing LLC. 2015;**5**(4): 047109

[2] Alamri SZ, Khan AA, Azeez M, Ellahi R. Effects of mass transfer on MHD second grade fluid towards stretching cylinder: A novel perspective of Cattaneo-Christov heat flux model. Physics Letters A. 2019;**383**(2–3): 276-281

[3] Hayat T, Khan MI, Farooq M, Alsaedi A, Waqas M, Yasmeen T. Impact of Cattaneo-Christov heat flux model in flow of variable thermal conductivity fluid over a variable thicked surface. International Journal of Heat and Mass Transfer. 2016;**99**:702-710

[4] Han S, Zheng L, Li C, Zhang X. Coupled flow and heat transfer in viscoelastic fluid with Cattaneo-Christov heat flux model. International Journal of Heat and Mass Transfer. 2009;**52**(21–22): 4675-4682

[5] Tibullo V, Zampoli V. A uniqueness result for the Cattaneo-Christov heat conduction model applied to incompressible fluids. Mechanics Research Communications. 2011;**38**:7779

[6] Oliveira LR, Silva OCA, Dantas NO, Filho EPB. Thermophycical properties of TiO2-PVA/water nanofluid. International Journal of Heat and Mass Transfer. 2017;**15**:795-808

[7] Waqas M, Hayat T, Farooq M, Shehzad SA, Alsaedi A. Cattaneo-Christov heat flux model for flow of variable thermal conductivity generalized Burgers fluid. Journal of Molecular Liquids. 2016;**220**(9–10): 642-648

[8] Hamilton RL, Crosser OK. Thermal conductivity of heterogeneous two component systems. Industrial and Engineering Chemistry Fundamentals. 1962;**1**(3):187-191

[9] Rangi RR, Ahmad N. Boundary layer flow past a stretching cylinder and heat transfer with variable thermal conductivity. Scientific Research. 2012; **3**(3):205-209

[10] Makinde OD, Nagendramma V, Raju CSK, Leelarathnam A. Effects of Cattaneo-Christov heat flux on Casson nanofluid flow past a stretching cylinder. International Journal of Heat and Mass Transfer. 2017;**378**:28-38

[11] Bibi S, Elahi Z, Shahzad A. Impacts of different shapes of nanoparticles on SiO_2 nanofuid flow and heat transfer in a liquid over a stretching sheet. Physica Scripta. 2020;**95**(11):1152178

Chapter 5

Nonlinear Dynamics Phenomenon in a Polydyne Cam with an Offset Flat Faced Follower Mechanism with Clearance

Louay S. Yousuf

Abstract

Nonlinear response of the follower motion is simulated at different cam speeds, different coefficient of restitution, and different internal distance of the follower guide from inside. The nonlinear response of the follower is employed to investigate the chaotic phenomenon in cam follower system in the presence of follower offset. The numerical results are done using SolidWorks software. The chaos phenomenon is detected using Poincare' maps with phase-plane portraits, the largest Lyapunov exponent parameter, and bifurcation diagram. The largest Lyapunov exponent has a maximum values when the follower offsets to the right, while the largest Lyapunov exponent has a minimum values when the follower offsets to the left. The chaotic phenomenon in cam follower system when the follower offsets to the left is more than the chaotic phenomenon when the follower offsets to the right.

Keywords: chaotic phenomenon, follower offset, Lyapunov exponent parameter, nonlinear response. Poincare' maps

1. Introduction

The proposed system can be found in windshield wiper on the front window of the car in which the rotary motion of the cam transforms into an oscillating motion. Yang et al. introduced the mathematical model to describe the separation, transient impact, and contact in cam follower system using oblique impact, [1]. They showed that the cam and the follower system kept permanent contact without the use of coefficient of restitution at low speeds for the cam. Yousuf studied the detachment between the cam and the follower using largest Lyapunov exponent parameter, power density function of Fast Fourier Transform (FFT), and Poincare' maps due to the nonlinear dynamics phenomenon of the follower. Nonlinear response of the follower displacement is calculated at different cam speeds, different coefficient of restitution, different contact conditions, and different internal distance of the follower guide from inside [2, 3]. Flores et al. used a nonsmooth dynamics approach to model the interaction of the colliding bodies using Coulomb's law for dry friction [4]. Lassaad et al. studied the effect of cam profile error on the nonlinear dynamics behavior of oscillating roller follower system by using a model with eight degrees of freedom of two nonlinear Hertzian contacts [5]. Li and Du used the coefficient of restitution as a main control parameter to analyze the periodic movement and the bifurcation region in Non-fixed constrained collision vibration system [6]. Wu et al. studied the influence of the joint clearance on the dynamic response of a planar mechanism with two driving links and prismatic pair clearance under variable input speeds [7]. They concluded that the largest Lyapunov exponents are dependent on the clearance size and the input speed. Chen et al. identified the chaos phenomenon of the 2-DOF nine-bar mechanism with a revolute clearance using the phase diagrams, the Poincaré portraits, and largest Lyapunov exponent parameter [8]. Bifurcation diagrams with changing clearance value, friction coefficient, and driving speed are drawn. The aim of this paper is to discuss the chaotic phenomenon of an offset follower through the use of impact coefficient of restitution at different follower guides' clearances and different cam speeds.

2. Numerical simulation

Follower displacement is calculated using SolidWorks software [9]. The follower moved with three degrees of freedom. Four values of the follower guide's from inside (I.D. = 16, 17, 18, 19 mm) at different cam speeds are used. The follower with the offset (O = 20, 30, 40, 50 mm) are chosen. The impact coefficient of restitution with the values (0.2, 0.3, and 0.4) is considered in the calculation of nonlinear response of the follower in the presence of follower offset. Cam follower mechanism is shown in **Figure 1**.

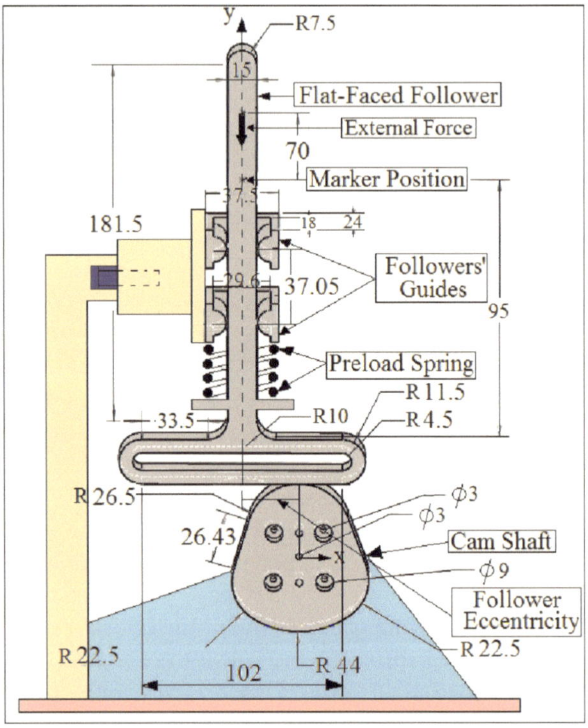

Figure 1.
Polydyne cam with an offset flat-faced follower.

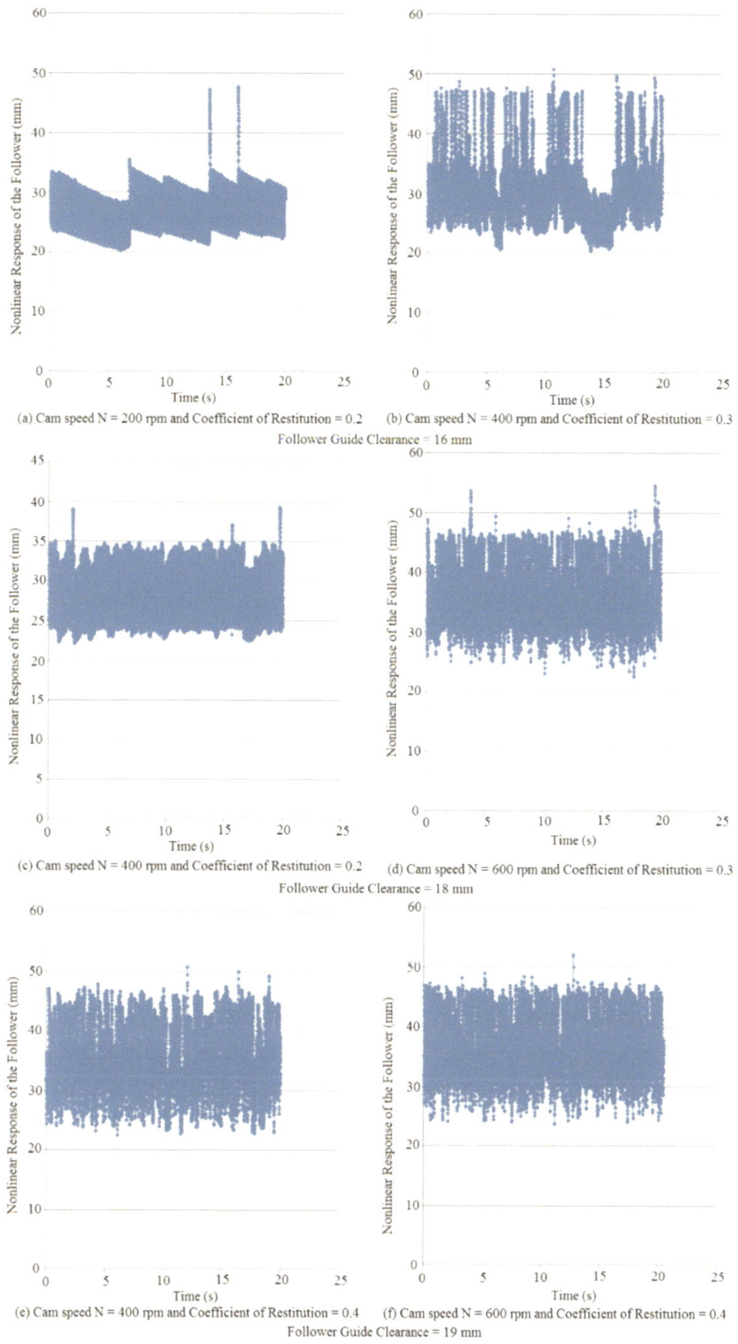

Figure 2.
Nonlinear response mapping when the follower offsets to the right (O = 10 mm).

The chaotic phenomenon in cam follower system is increased with the increasing of impact coefficient of restitution in which the impact will happen due the loss in potential energy of the follower and due to the increase in follower guide clearance value. **Figures 2** and **3** show the mapping of nonlinear response of the follower at

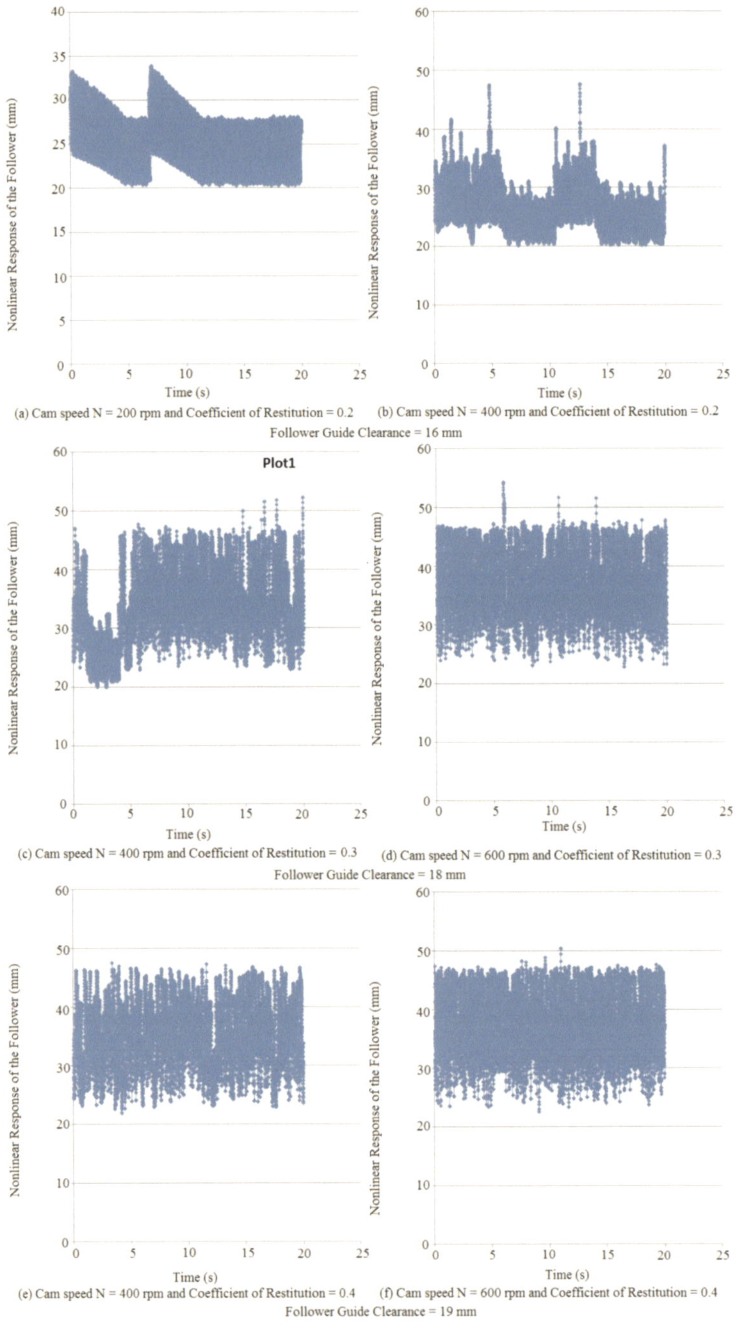

Figure 3.
Nonlinear response mapping when the follower offsets to the left (O = 10 mm).

different cam speeds, different follower guides' clearances, and different impact coefficient of restitution when the follower offsets to the right and left respectively (O = 10 mm). The nonlinear response of the follower is periodic as shown in **Figure 2a** and both the cam and the follower are in permanent contact. The follower lost the contact with the cam at time (t = 13.58 s) and (t = 15.99 s) at detachment height

(26.98 mm) and (27.43 mm) respectively. Due to the coefficient of restitution, the follower keep bouncing from the cam from (t = 0.36 s) to (t = 5.658 s) while the follower will regain energy and keep permanent contact with the cam for the period from (t = 9.208 s) to (t = 10.11 s) which is having a periodic motion as illustrated in **Figure 2b**. The chaotic motion is shown in **Figure 2c–f** which increased with the increasing of follower guides' clearances, cam speeds, and coefficient of restitution. There is an intangible impact when the coefficient of restitution (0.2) and the dissipation in potential energy is occurred due to sliding while the contact is still valid between the cam and the follower, as shown in **Figure 3a**. The periodic and chaotic motion is together shown in **Figure 3b** and **c**. The periodic motion is shown from the period (t = 6.1 s) to (t = 10.26 s) and from the period (t = 14.14 s) to (t = 19.55 s) as shown in **Figure 3b** while the periodic motion begins from the period (t = 1.264 s) to (t = 3.808 s) as shown in **Figure 3c**. The chaotic motion is shown in **Figure 3d–f**.

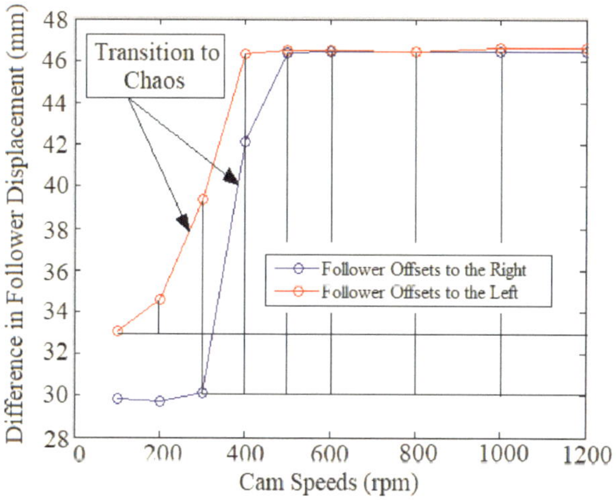

Figure 4.
Bifurcation diagram against cam speeds.

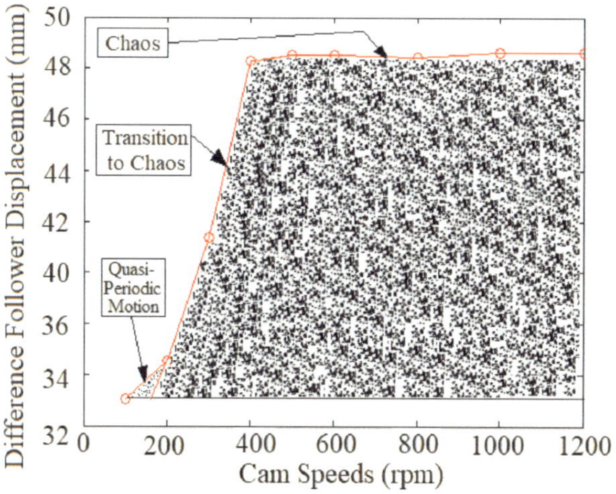

Figure 5.
Bifurcation diagram when the follower offsets to the left (O = 50 mm).

3. Bifurcation diagram

The contrast in angular displacement for the cam and the follower is used in the calculation of bifurcation diagram [10, 11]. **Figure 4** is built at the follower guide's from inside (I.D. = 19 mm) when the follower offsets to the right and left (O = 50 mm).

The periodic motion is shown in **Figure 4** in which it has the blue trend at cam speeds (N = 100–300 rpm) while the quasi-periodic motion of the follower has red trend at cam speeds (N = 100 rpm). The transition to chaos for the system when the follower offsets to the left is grown faster than the system when the follower offsets to the right as indicated in **Figures 5** and **6**. It can be concluded that the transition to chaos is incremented with the increment in cam speeds.

4. Lyapunov exponent parameter

Local Lyapunov exponent parameter is used to detect the chaotic phenomenon of nonlinear response of the follower attractor. Positive Lyapunov exponent refers to chaotic phenomenon while negative Lyapunov exponent indicates to periodic motion [12]. **Figure 7** shows the local Lyapunov exponent against number of points when the follower offsets to the right (O = 10 mm) at coefficient of restitution (0.2), cam speed (N = 200 rpm), and follower guide's clearance (16 mm). In this figure there are positive and negative local Lyapunov exponent in which the negative local Lyapunov exponent represents to the steady state while the positive local Lyapunov exponent reflects to the transient state. Each value of local Lyapunov exponent has a value of embedding dimension [13].

5. Poincare' maps with phase-plane portraits

The contact status of the follower is detected using Poincare' map at high and low speeds [14]. Moreover, the quantity of the black dots in Poincare' maps detects the

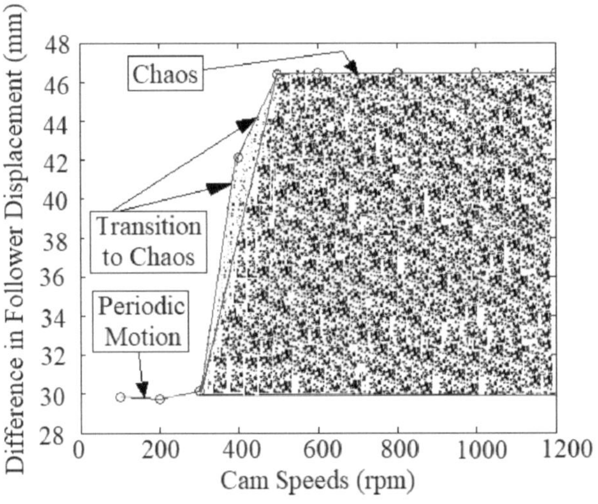

Figure 6.
Bifurcation diagram when the follower offsets to the right (O = 50 mm).

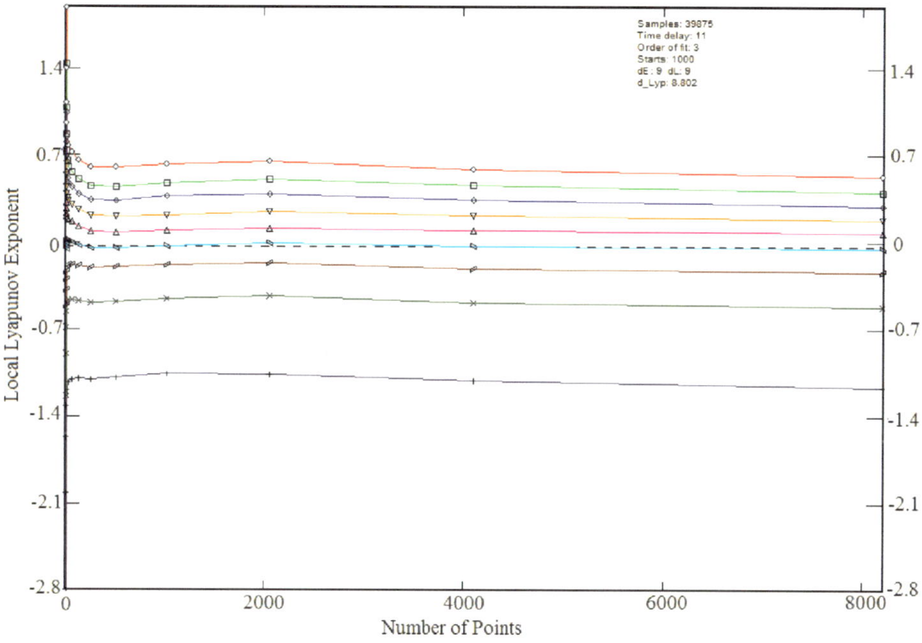

Figure 7.
Local Lyapunov exponent when the follower offsets to the right (O = 10 mm) at cam speed (N = 200 rpm), coefficient of restitution (0.2), and follower guide's clearance (16 mm).

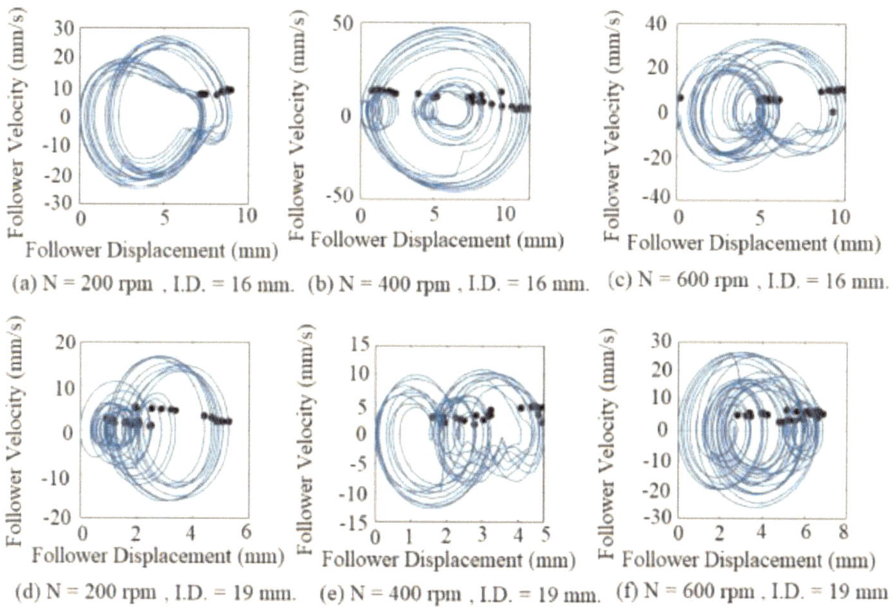

Figure 8.
Phase portrait of chaotic attractor when the follower offsets to the left (O = 20 mm).

chaotic analysis in follower movement when the follower has detached from the cam. The system in **Figure 8** has smooth orbit of follower displacement at (I.D. = 16 mm and 19 mm) when the follower offsets to the left (O = 20 mm) at diverse cam speeds.

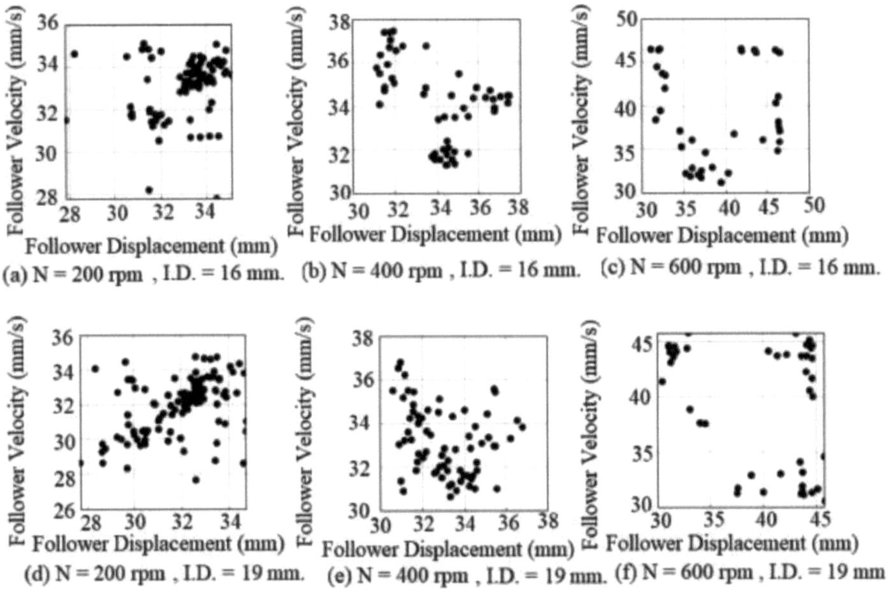

Figure 9.
Poincare' maps of chaotic attractor when the follower offsets to the left (O = 20 mm).

The follower displacement is repeated itself based on the single black dots in phase-plane orbit. The chaotic analysis is detected based on the multi black dots in phase-plane orbit at (I.D. = 16 mm and 19 mm) and diverse cam speeds as shown in **Figure 9**. SolidWorks software is used in the simulation.

6. Follower displacement

Figures 10 and **11** show the follower linear displacement against the time at different cam angular speeds when the camshaft offsets to the left (O = 40 mm) and to the right (O = 50 mm) at (I.D. = 17 mm) respectively. The follower stays in permanent contact when the cam starts spinning at (N = 200 rpm) and (N = 400 rpm), while the follower starts detaching from the cam at (N = 1000 rpm) as shown in **Figure 10**. The follower also starts jumping a little bit higher from the cam at (N = 800 rpm) as shown in **Figure 11**.

MATLAB Code:

The code algorithm of phase-plane diagram and Poincare' map are added at the end of this chapter. The code is done using MATLAB software and as in below:

```
clear; clc; close all
SignalName = '100rpm.dat';
signal = load(SignalName);
signal = signal - min(signal);
% Poincare map
% original D=signal; % Read data
[x1max,t1max] = findpeaks(D(:,1));
```

DOI: http://dx.doi.org/10.5772/intechopen.106179

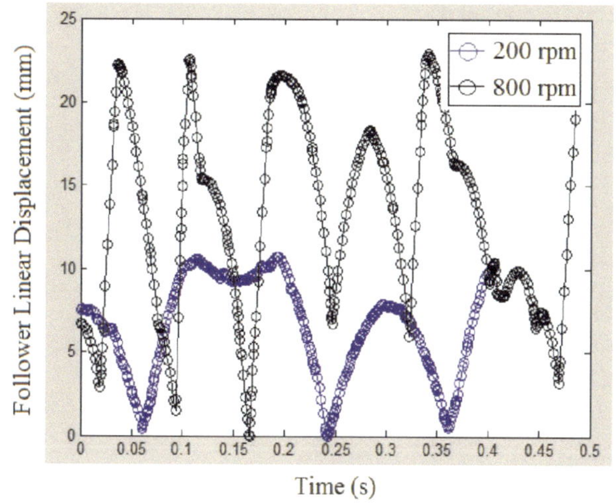

Figure 10.
Follower displacement against time when the follower offsets to the right (O = 50 mm) at various cam speeds.

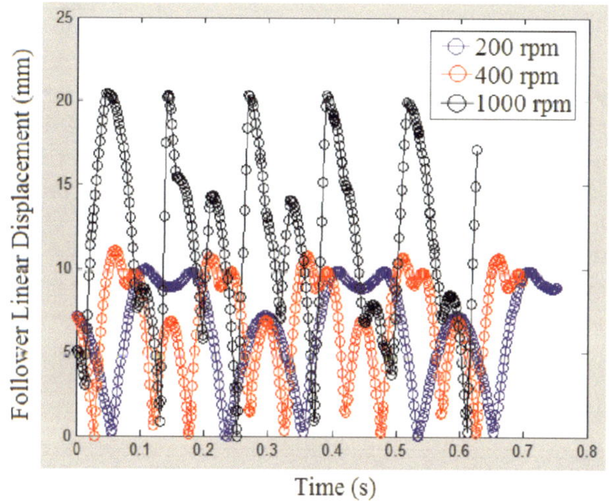

Figure 11.
Follower displacement against time when the follower offsets to the left (O = 40 mm) at various cam speeds.

```
    Nmax = length(x1max); figure(1)
    subplot(1,2,1)
    for i=1:Nmax-1 plot(x1max(i),x1max(i+1),'ko','MarkerSize',5,'MarkerFa-
ceColor','k')
    hold on
    axis square
    xlabel('x_{max} '),ylabel('next x_{max}')
    grid on
    end
    %title('n = 100 rpm c = 1.5 mm') SignalName = '100rpmc2.dat';
    signal = load(SignalName);
    D = signal; % Read data
```

```
[x1max,t1max] = findpeaks(D(:,1));
Nmax = length(x1max); subplot(1,2,2)
for i=1:Nmax-1 plot(x1max(i),x1max(i+1),'ko','MarkerSize',5,'MarkerFa-
ceColor','k')
hold on
axis square
xlabel('x_{max} '),ylabel('next x_{max}')
grid on
end
%title('n = 100 rpm c = 2 mm') figure(2)
aa = load('100rpm.dat');
aa = aa - min(aa);
plot(aa, gradient(aa));
```

7. Conclusions

In this article the chaotic motion of the follower response is considered in the presence of impact coefficient of restitution using SolidWorks program. The chaotic motion of the follower response is occurred due to the increase in cam speeds, follower's offsets, follower guides' clearances and impact coefficient of restitution. The value of Lyapunov exponent is increased with the increasing of embedding dimensions values. The positive local Lyapunov exponent depicts the transient state in nonlinear response of the follower in the presence of impact coefficient of restitution. Negative local Lyapunov exponent refers to the steady state in the follower motion. Some of the nonlinear response of the follower has periodic and chaotic motions at different time periods. The quantity of the black dots in Poincare' maps detects the chaotic analysis in follower movement when the follower has detached from the cam.

Acknowledgements

The author wants to thank the editor and the reviewers for their helpful suggestion.

Conflict of interest

The author declares that he has no conflict of interest.

Author details

Louay S. Yousuf
Department of Mechanical Engineering, San Diego State University, San Diego, CA, USA

*Address all correspondence to: louaysabah79@yahoo.com

References

[1] Yang YF, Lu Y, Jiang TD, Lu N. Modeling and nonlinear response of the cam-follower oblique impact system. In: Discrete Dynamics in Nature and Society. New York, USA: Hindawi Publisher; 2016

[2] Yousuf LS. Detachment detection in cam follower system due to nonlinear dynamics phenomenon. Journal of Machine, Special Issue: Dynamics Analysis of Multibody Mechanical Systems. 2021;**9**(12):349

[3] Yousuf LS. Nonlinear dynamics phenomenon detection in a polydyne cam with an offset flat-faced follower mechanism using multi shocks absorbers systems. Journal of Applications in Engineering Science. 2022:100086

[4] Flores P, Leine R, Glocked C. Application of the nonsmooth dynamics approach to model and analysis of the contact-impact events in cam-follower systems. Nonlinear Dynamics. 2012; **69**(4):2117-2133

[5] Lassaad W, Mohamed T, Yassine D, Fakher C, Mohamed H. Nonlinear dynamic behaviour of a cam mechanism with oscillating roller follower. Frontiers of Mechanical Engineering. 2013;**8**(2): 127-136

[6] Li Z, Du Y. Interval of restitution coefficient for chattering in impact damper. Journal of Low Frequency Noise, Vibration and Active Control. 2021;**41**(2):432-450

[7] Wu L, Marghitu DB, Zhao J. Nonlinear dynamics response of a planar mechanism with two driving links and prismatic pair clearance. Mathematical Problems in Engineering Journal. 2017; **2017**:1-12

[8] Chen X, Jiang S, Deng Y, Wang Q. Nonlinear dynamics and analysis of a planar multilink complex mechanism with clearance. Journal of Shock and Vibration. 2018;**2018**:1-17

[9] Planchard D. SolidWorks 2016 Reference Guide: A Comprehensive Reference Guide with over 250 Standalone Tutorials. Mission, Kansas, USA: Sdc Publisher; 2015

[10] Yousuf LS. Investigation of chaos in a polydyne cam with flat-faced follower mechanism. Journal of King Saudi University and Engineering Science. 2020, 2020;**33**(7):507-516

[11] Yousuf LS. Non-periodic motion reduction in globoidal cam with roller follower mechanism. Proceedings of the Institution of Mechanical Engineers. 2021;**236**(6):2714-2727

[12] Terrier P, Reynard F. Maximum Lyapunov exponent revisited: Long-term attractor divergence of gait dynamics is highly sensitive to the noise structure of stride intervals. Gait & Posture. 2018;**66**: 236-241

[13] Hussain VS, Spano ML, Lockhart TE. Effect of data length on time delay and embedding dimension for calculating the Lyapunov exponent in walking. Journal of the Royal Society Interface. 2020; **17**(168):20200311

[14] Abu-Mahfouz I. Experimental investigation of non-linear behavior of a cam-follower mechanism. In: SEM Annual Conference and Exposition on Experimental Applied Mechanics. Springfield, MA, USA: SEM Publisher; 2005. pp. 1930-1936

Chapter 6

Decision Fusion for Large-Scale Sensor Networks with Nonideal Channels

Yiwei Liao, Xiaojing Shen, Junfeng Wang and Yunmin Zhu

Abstract

Since there has been an increasing interest in the areas of Internet of Things (IoT) and artificial intelligence that often deals with a large number of sensors, this chapter investigates the decision fusion problem for large-scale sensor networks. Due to unavoidable transmission channel interference, we consider sensor networks with nonideal channels that are prone to errors. When the fusion rule is fixed, we present the necessary condition for the optimal sensor rules that minimize the Monte Carlo cost function. For the K-out-of-L fusion rule chosen very often in practice, we analytically derive the optimal sensor rules. For general fusion rules, a Monte Carlo Gauss-Seidel optimization algorithm is developed to search for the optimal sensor rules. The complexity of the new algorithm is of the order of $O(LN)$ compared with $O(LN^L)$ of the previous algorithm that was based on Riemann sum approximation, where L is the number of sensors and N is the number of samples. Thus, the proposed method allows us to design the decision fusion rule for large-scale sensor networks. Moreover, the algorithm is generalized to simultaneously search for the optimal sensor rules and the optimal fusion rule. Finally, numerical examples show the effectiveness of the new algorithms for large-scale sensor networks with nonideal channels.

Keywords: decision fusion, multisensor detection, nonideal channels, Monte Carlo method, importance sampling

1. Introduction

Distributed detection has been an active research area in the past decades [1–7]. It involves the design of decision rules for the sensors[1] and the fusion rule [8]. Early work on distributed detection mainly focused on conditionally independent sensor observations, such as [2, 4, 9, 10], and the resulting optimal sensor decision rules, as well as the fusion rule, were likelihood ratio tests (LRTs). Details on distributed detection with conditionally independent sensor observations can be seen in [1, 6, 7] and references therein.

1 In the rest of the paper, the term "sensor rules" refers to the "decision rules at the sensors."

In this chapter, we focus on conditionally dependent observations in sensor networks. In [5], the computational difficulty of obtaining the optimal sensor rules was shown by a rigorous mathematical approach. Some early progress was made on the derivation of sensor rules for the dependent observation case such as in [11–15]. More recently, a hierarchical conditional independence model was provided that was applicable to some specific classes of multisensor detection problems with dependent observations [16]. Copula-based distributed decision fusion methods have been proposed to deal with dependent observations in sensor networks, such as [17–19] and references therein. Given a fusion rule, Monte Carlo methods were proposed to reduce the computational complexity of deriving sensor decision rules with ideal channels in [20, 21], and the optimal sensor rules were obtained analytically for the K-out-of-L fusion rule in [20].

Some works on the derivation of optimal fusion rules can be seen in [15, 22–24]. For some specific parallel network decision systems, a unified fusion rule was presented in [15]. Some further results on the problem are available in [25, 26]. In [27], the authors provided methods that search for the sensor rules and the fusion rule simultaneously by combining the methods of [2] and [15] in order to attain near-optimal system performance.

The works discussed thus far assumed the availability of ideal channels in sensor networks. However, channel errors between the sensors and the fusion center are omnipresent in practical multisensor detection networks, and, therefore, studies on multisensor detection in the presence of nonideal channels have attracted some recent interest, such as in [8, 28–33]. Under the Neyman-Pearson criterion, the design of sensor rules in the presence of nonideal channels was addressed in [32]. The parallel fusion structure was extended by incorporating the fading channel layer and two alternative fusion schemes were presented based on fixed sensor rules in [28]. It was shown that the optimal sensor decision rule that minimizes the error probability at the fusion center is equivalent to a local LRT for independent sensor observations in [29]. Under Neyman-Pearson and Bayesian criteria, the work was generalized to dependent and noisy channels, respectively, in [8]. In [31], the authors considered the optimal sensor rules with channel errors *via* Riemann sum approximation under a given fusion rule for general dependent sensor observations. Although the method based on the Riemann sum approximation has been developed for dependent observations with channel errors, it is too computationally expensive to be of practical use in large-scale sensor networks.

In this chapter, a Monte Carlo importance sampling method is provided to reduce the computational complexity of multisensor detection fusion with channel errors. Based on the strong law of large numbers, the Bayesian cost function is approximated by its empirical average through the Monte Carlo importance sampling method. The main contributions of this chapter are listed below:

1. When the fusion rule is fixed, we derive a necessary condition for the optimal sensor rules that minimize the approximated Bayesian cost function. A Monte Carlo Gauss-Seidel optimization algorithm is developed and it is shown to be finitely convergent. The complexity of the new algorithm is shown to be of the order of $O(LN)$ compared with $O(LN^L)$ of the previous algorithm based on the Riemann sum approximation.

2. When the fusion rule is the K-out-of-L rule, we prove that there exists an analytical form for the optimal sensor rules in the presence of nonideal channels. Thus, the proposed method allows us to design decision rules for large-scale sensor networks.

3. The Monte Carlo Gauss-Seidel optimization algorithm is extended to simultaneously search for the optimal sensor rules and the optimal fusion rule.

Numerical examples show the effectiveness of the new algorithms for large-scale sensor networks with dependent observations and channel errors.

The rest of this chapter is organized as follows: In Section 2, the parallel binary Bayesian detection network with channel errors is formulated and the Monte Carlo cost function is introduced. In Section 3, the necessary condition for the optimal sensor rules is presented. For the K-out-of-L fusion rule, the analytical form for the optimal sensor rules is provided. In Section 4, the Monte Carlo Gauss-Seidel iterative algorithm and its convergence analysis are presented. The extension to search for the optimal sensor rules and the optimal fusion rule are simultaneously described in Section 5. Simulation results are provided in Section 6. Conclusions are contained in Section 7.

2. Preliminaries

2.1 Problem formulation

The L-sensor parallel Bayesian detection network structure with two hypotheses H_0 and H_1 in the presence of nonideal channels is considered (see **Figure 1**). Assume that $y_1, y_2, \dots, y_L$ are sensor observations and the jth sensor compresses the n_j-dimension vector observation y_j to one bit: $I_j\left(y_j\right) : \mathbb{R}^{n_j} \to \{0, 1\}, j = 1, \dots, L$. For notational convenience, $n_j = 1$ in the following description. The L sensors transmit the compressed data to the fusion center and the fusion center makes the decision between H_0 and H_1. Since external interference and internal errors may occur, the channels are not reliable and the fusion center may not correctly receive the symbol I_j sent by the jth sensor. Let I_j^0 denote the received bit by the fusion center for $j = 1, 2, \dots, L$. Generally speaking, I_j^0 may not be equal to I_j. The definition and assumptions on channel errors (see e.g., [29, 31]) are summarized below:

Definition 1: The channel errors between the jth sensor and the fusion center are described as $P_j^{ce1} = P\left(I_j^0 = 0 | I_j = 1\right)$ and $P_j^{ce0} = P\left(I_j^0 = 1 | I_j = 0\right)$ for $j = 1, 2, \dots, L$, where P_j^{ce1} is the probability of channel error when the jth sensor sends 1 but the fusion center receives 0, and P_j^{ce0} is the probability of channel error when the jth sensor sends 0 but the fusion center receives 1.

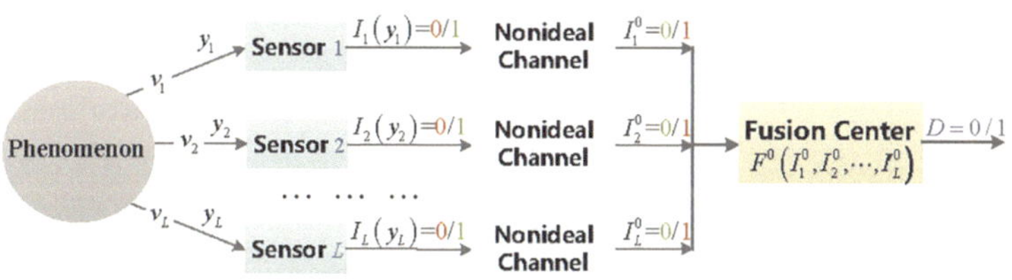

Figure 1.
The L-sensor parallel binary Bayesian detection network structure in the presence of nonideal channels.

Assumption 1: The probabilities of channel error are statistically independent of the hypotheses, namely $P\left(I_j^0|I_j, H_\nu\right) = P\left(I_j^0|I_j\right)$, $\nu = 0, 1$.

Remark 1: Assumption 1 is due to the hierarchical structure based on the Markov property (see [29]).

Assumption 2: The channels that connect the sensors to the fusion center are independent, i.e., $P(I_1^0, I_2^0, \ldots, I_L^0|I_1, I_2, \ldots, I_L) = \prod_{j=1}^{L} P\left(I_j^0|I_j\right)$.

We consider the parallel binary Bayesian detection network with nonideal channels that is built on the above definition and assumptions. The final decision is made by the fusion center based on the received binary bits $(I_1^0, I_2^0, \ldots, I_L^0)$ from the L sensors. From the definition of a general Bayesian cost function given in [25], the L-sensor binary Bayesian cost function with channel errors at the fusion center can be written as follows:

$$\begin{aligned} C(I_1^0(y_1), \ldots, I_L^0(y_L); F^0) = C_{00}P_0P(F^0 = 0|H_0) + C_{01}P_1P(F^0 = 0|H_1) \\ + C_{10}P_0P(F^0 = 1|H_0) + C_{11}P_1P(F^0 = 1|H_1) \end{aligned} \tag{1}$$

$$= c + aP(F^0 = 0|H_1) - bP(F^0 = 0|H_0), \tag{2}$$

where $C_{\alpha\beta}$, $\alpha, \beta = 0, 1$ are suitable cost coefficients, P_0 and P_1 are the prior probabilities for the hypotheses H_0 and H_1, respectively, F^0 is the fusion rule, and $P(F^0 = \mu|H_\nu)$, $\mu, \nu = 0, 1$ denotes the conditional probability of the event that the fusion center decides in favor of hypothesis μ when the real hypothesis is H_ν. The cost function (1) is simplified to (2) by defining $c = C_{10}P_0 + C_{11}P_1$, $a = P_1(C_{01} - C_{11})$, $b = P_0(C_{10} - C_{00})$. F^0 is actually a function of the disjoint set of all possible binary messages $(I_1^0, I_2^0, \ldots, I_L^0)$. The received decisions are divided into two sets denoted as H_0^0 and H_1^0 which are given by

$$H_0^0 = \left\{\left(u_1^0, u_2^0, \ldots, u_L^0\right) : F^0\left(\left(I_1^0, I_2^0, \ldots, I_L^0\right)\right) = 0, I_j^0 = u_j^0, u_j^0 = 0/1, j = 1, \ldots, L\right\};$$

$$H_1^0 = \left\{\left(u_1^0, u_2^0, \ldots, u_L^0\right) : F^0\left(\left(I_1^0, I_2^0, \ldots, I_L^0\right)\right) = 1, I_j^0 = u_j^0, u_j^0 = 0/1, j = 1, \ldots, L\right\}.$$

Obviously, $H^0 = \left\{\left(u_1^0, u_2^0, \ldots, u_L^0\right) : I_j^0 = u_j^0, u_j^0 = 0/1, j = 1, \ldots, L\right\} = H_0^0 \cup H_1^0$. For any binary decisions $\left(I_1^0, I_2^0, \ldots, I_L^0\right)$ received by the fusion center, the original sensor decision bits before transmission are $(I_1, I_2, \ldots, I_L)$ and they consist of the set $H = \{(u_1, u_2, \ldots, u_L) : I_j = u_j, u_j = 0/1, j = 1, \ldots, L\}$. Therefore, based on the law of total probability, the conditional probability formula, and Assumption 1:

$$P(F^0 = 0|H_\nu) = \sum_{s^0 \in H_0^0} P(D^0|H_\nu) = \sum_{s^0 \in H_0^0} \sum_{s \in H} P(D^0|D)P(D|H_\nu), \tag{3}$$

where $D^0 = \left(I_1^0, I_2^0, \ldots, I_L^0\right)$, $s^0 = \left(s^0(1), \ldots, s^0(L)\right)$, $I_j^0 = s^0(j)$, and $s^0(j) = 0/1$ is a specific value of I_j^0; in the same way, $D = (I_1, I_2, \ldots, I_L)$, $s = (s(1), \ldots, s(L))$, $I_j = s(j)$, and $s(j) = 0/1$ is a specific value of I_j. Strictly speaking, we should use $P(D^0 = s^0|H_\nu)$ to represent $P(D^0|H_\nu)$ and we use the latter for notational simplicity. It is similar to $P(D|H_\nu)$. Based on Assumption 2:

$$P(D^0|D) = \prod_{j=1}^{L} P\left(I_j^0|I_j\right), \tag{4}$$

where for any $1 \leq j \leq L$

$$P\left(I_j^0|I_j\right) = \left(1 - P_j^{ce0}\right)\left(1 - I_j^0\right)(1 - I_j) + P_j^{ce0} I_j^0 (1 - I_j) + \left(1 - P_j^{ce1}\right) I_j^0 I_j + P_j^{ce1}\left(1 - I_j^0\right) I_j. \tag{5}$$

Thus, the cost function (2) becomes

$$C\left(I_1^0(y_1), \ldots, I_L^0(y_L); F^0\right) = c + \sum_{s^0 \in H_0^0} \sum_{s \in H} P(D^0|D)[aP(D|H_1) - bP(D|H_0)] \tag{6}$$

$$\triangleq C\left(I_1(y_1), \ldots, I_L(y_L); F^0; P^{ce0}, P^{ce1}\right), \tag{7}$$

where $P^{ce0} = \left(P_1^{ce0}, \ldots, P_L^{ce0}\right)$, $P^{ce1} = \left(P_1^{ce1}, \ldots, P_L^{ce1}\right)$. Hence, the cost function now becomes a function of the sensor rules $(I_1, \ldots, I_L)$, the probabilities of channel errors P^{ce0}, P^{ce1}, and the fusion rule F^0. The goal of this chapter is to optimize the sensor rules and the fusion rule so as to minimize the cost function with known probabilities of channel errors.

We rewrite $aP(D|H_1) - bP(D|H_0)$ as follows:

$$\begin{aligned} aPD|H_1 - bPD|H_0 &= \int_{\Omega_s} apy_1, \ldots, y_L|H_1 - bpy_1, \ldots, y_L|H_0 dy_1 \cdots dy_L \\ &= \int I_{\Omega_s}\left[apy_1, \ldots, y_L|H_1 - bpy_1, \ldots, y_L|H_0\right] dy_1 \cdots dy_L, \end{aligned} \tag{8}$$

where $\Omega_s = \left\{(y_1, \ldots, y_L) : I_1(y_1) = s(1), \ldots, I_L(y_L) = s(L)\right\}$, I_{Ω_s} is an indicator function on Ω_s, and the region of integration in (8) is the full space. Assume that $p(y_1, y_2, \ldots, y_L|H_\nu)$, $\nu = 0, 1$ (or $p(y|H_\nu)$) are the known conditional joint probability density functions. If not, we can learn the joint probability density functions from training data using copula functions (see, e.g., [17]). Note that $I_1(y_1), \ldots, I_L(y_L)$ are indicator functions and $s(j) = 0/1, j = 1, \ldots, L$,

$$\begin{aligned} I_{\Omega_s} &= I_{\left\{(y_1, \ldots, y_L): I_1(y_1) = s(1), \ldots, I_L(y_L) = s(L)\right\}} \\ &= I_{\left\{y_1: I_1(y_1) = s(1)\right\}} \cdots I_{\left\{y_L: I_L(y_L) = s(L)\right\}} \\ &= [(1 - I_1)(1 - s(1)) + I_1 s(1)] \cdots [(1 - I_L)(1 - s(L)) + I_L s(L)]. \end{aligned} \tag{9}$$

For simplicity, denote $Q_j(I_j) = (1 - I_j)(1 - s(j)) + I_j s(j)$. Substituting (8) into (6),

$$\begin{aligned} &C\left(I_1(y_1), \ldots, I_L(y_L); F^0; P^{ce0}, P^{ce1}\right) \\ &= c + \sum_{s^0 \in H_0^0} \sum_{s \in H} P(D^0|D) \cdot \int Q_1(I_1) \cdots Q_L(I_L)[ap(y|H_1) - bp(y|H_0)] dy = c + \int P_{H_0^0} \hat{L}(y) dy, \end{aligned} \tag{10}$$

where $P_{H_0^0} = \sum_{s^0 \in H_0^0} \sum_{s \in H} P(D^0|D) Q_1(I_1) \cdots Q_L(I_L)$ and $\hat{L}(y) = ap(y|H_1) - bp(y|H_0)$. Note that from the definition of H_0^0, H_1^0, H^0, and H, we have

$$P_{H_0^0} = \sum_{s^0 \in H^0} \left[1 - F^0\left(D^0\right)\right] \sum_{s \in H} P\left(D^0 | D\right) Q_1(I_1) \cdots Q_L(I_L)$$
$$= \sum_{k'=1}^{2^L} \sum_{k=1}^{2^L} \left[1 - F^0(s_{k'})\right] P(s_{k'} | s_k) \cdot \prod_{j=1}^{L} \left\{\left[1 - I_j\left(y_j\right)\right]\left[1 - s_k(j)\right] + I_j\left(y_j\right) s_k(j)\right\}, \tag{11}$$

where $s_{k'}$ is the element of H^0 and s_k is the element of H. $F^0\left(D^0\right) = 0/1$ is used in the first equality. The second equality holds since there are 2^L elements in both H and H^0.

2.2 Monte Carlo cost function

An essential difficulty of the Bayesian cost function (10) is the required high dimensional integration when dealing with large-scale sensor networks. Monte Carlo importance sampling is an attractive method to deal with this problem. In this subsection, we approximate the Bayesian cost function (10) by the Monte Carlo importance sampling method (see, e.g., [34, 35]). According to (10),

$$C\left(I_1(y_1), \ldots, I_L\left(y_L\right); F^0; P^{ce0}, P^{ce1}\right)$$
$$= c + \int P_{H_0^0}\left(I_1(y_1), \ldots, I_L\left(y_L\right); F^0; P^{ce0}, P^{ce1}\right) \frac{\hat{L}(y)}{g(y)} \cdot g(y) dy \tag{12}$$

$$= \mathbb{E}_g \frac{P_{H_0^0}(Y)\hat{L}(Y)}{g(Y)} + c, \tag{13}$$

where $y = \left(y_1, y_2, \ldots, y_L\right)$, and $g(y)$ is a given importance sampling density such that (12) is well-defined (i.e., $g(y) > 0$). In (13), the expectation is taken with respect to the importance sampling density g. Consequently, assume that N samples $Y_1, \ldots, Y_N$ are generated from the density g, that is, $Y \sim g(y)$, where $Y_i = \left[Y_{i1}, Y_{i2}, \ldots, Y_{iL}\right]$. Then

$$C\left(I_1(y_1), \ldots, I_L\left(y_L\right); F^0; P^{ce0}, P^{ce1}\right) \approx \frac{1}{N} \sum_{i=1}^{N} \frac{P_{H_0^0}(Y_{i1}, Y_{i2}, \ldots, Y_{iL})\hat{L}(Y_i)}{g(Y_i)} + c \tag{14}$$

$$\triangleq C_{MC}\left(I_1(y_1), \ldots, I_L\left(y_L\right); F^0; P^{ce0}, P^{ce1}, N\right). \tag{15}$$

Based on the strong law of large numbers, the expectation (13) can be approximated by the empirical average (14). Denote (14), namely the Monte Carlo cost function, as $C_{MC}\left(I_1(y_1), \ldots, I_L\left(y_L\right); F^0; P^{ce0}, P^{ce1}, N\right)$. The optimal importance sampling density is $g\left(y_1, y_2, \ldots, y_L\right) \propto \left|P_{H_0^0}\hat{L}\left(y_1, y_2, \ldots, y_L\right)\right|$ (see, e.g., [34, 35]).

The initial goal is to minimize the Bayesian cost function (10). Instead, we can minimize the Monte Carlo cost function (15) by selecting a set of optimal sensor rules $I_1(y_1), I_2(y_2), \ldots, I_L\left(y_L\right)$ and an optimal fusion rule F^0. In this manner, the high-dimensional integration problem is converted to a problem where we need to deal with the single summation objective function for large-scale sensor networks. Thus, for dependent observations with channel errors, the computational complexity is reduced significantly by the Monte Carlo importance sampling method. In the following sections, we assume that the samples drawn from the importance sampling density are fixed so that $C_{MC}(I_1, \ldots, I_L; F^0; P^{ce0}, P^{ce1}, N)$ does not have any randomness, since only deterministic decision rules are considered in this chapter.

3. A necessary condition for the optimal sensor rules

In this section, when the fusion rule is fixed, we derive a necessary condition for the optimal sensor rules that minimize the Monte Carlo cost function. First, we need some equivalent transformations for $P_{H_0^0}$. Then based on the transformations, the necessary condition can be obtained. At the same time, an analytical result is obtained when the fusion rule is the K-out-of-L rule.

3.1 Necessary condition

First, we need some equivalent transformations for $P_{H_0^0}$.

Lemma 1 $P_{H_0^0}$ can be rewritten as follows:

$$P_{H_0^0} \triangleq \left[1 - I_j\left(y_j\right)\right] P_{j1}\left(I_1(y_1), \ldots, I_{j-1}\left(y_{j-1}\right), I_{j+1}\left(y_{j+1}\right), \ldots, I_L\left(y_L\right); F^0; P^{ce0}, P^{ce1}\right)$$
$$+ P_{j2}\left(I_1(y_1), \ldots, I_{j-1}\left(y_{j-1}\right), I_{j+1}\left(y_{j+1}\right), \ldots, I_L\left(y_L\right); F^0; P^{ce0}, P^{ce1}\right), \tag{16}$$

where for $j = 1, 2, \ldots, L$,

$$P_{j1}(\cdot) \triangleq \sum_{k'=1}^{2^L} [1 - F(s_{k'})]\left(1 - P_j^{ce0} - P_j^{ce1}\right)(1 - 2s_{k'}(j)) P_{m \neq j}, \tag{17}$$

$$P_{j2}(\cdot) \triangleq \sum_{k'=1}^{2^L} [1 - F(s_{k'})]\left[s_{k'}(j) + P_j^{ce1}(1 - 2s_{k'}(j))\right] P_{m \neq j}, \tag{18}$$

$$P_{m \neq j} \triangleq \prod_{m,\, m \neq j}^{L} \{(1 - P_m^{ce0})(1 - s_{k'}(m))(1 - I_m) + P_m^{ce0} s_{k'}(m)(1 - I_m)$$
$$+ (1 - P_m^{ce1}) s_{k'}(m) I_m + P_m^{ce1}(1 - s_{k'}(m)) I_m\}. \tag{19}$$

Proof: If $s_k(m) = I_m\left(y_m\right)$ for all $m = 1, \ldots, L$, then the continued product term $\prod_{m=1}^{L}\{[1 - I_m\left(y_m\right)][1 - s_k(m)] + I_m\left(y_m\right) s_k(m)\} = 1$ in $P_{H_0^0}$. Otherwise, it is 0. Thus, $P_{H_0^0}$ can be rewritten as $P_{H_0^0} = \sum_{k'=1}^{2^L}[1 - F(s_{k'})] P(s_{k'}|(I_1, I_2, \ldots, I_L))$, where the terms that equal zero are omitted and $P(s_{k'}|(I_1, I_2, \ldots, I_L)) = \prod_{j=1}^{L} P\left(s_{k'}(j)|I_j\right)$. Recalling the conditional probability formula (5), we rewrite $P\left(s_{k'}(j)|I_j\right)$ as $P\left(s_{k'}(j)|I_j\right) = \left[1 - I_j\right]\left(1 - P_j^{ce0} - P_j^{ce1}\right)(1 - 2s_{k'}(j)) + s_{k'}(j) + P_j^{ce1}(1 - 2s_{k'}(j))$. Based on these transformations, $P_{H_0^0}$ can be decomposed as (16).

Remark 2: Note that $P_{j1}(\cdot)$ and $P_{j2}(\cdot)$ are both independent of $I_j\left(y_j\right)$ for $j = 1, \ldots, L$. In addition, they can also be applied in the Riemann sum approximation (see, e.g., [31]).

Compared with [36], the sum of 2^L terms about s_k is eliminated and it greatly reduces the computational time. In addition, the expression for $P_{j1}(\cdot)$ given in (17) is also a key equation in the following results:

Substituting the transformations (16) into (15), we obtain

$$C_{MC}(I_1(y_1), \dots, I_L(y_L); F^0; P^{ce0}, P^{ce1}, N)$$
$$= c + \frac{1}{N}\sum_{i=1}^{N}\{[1 - I_j(Y_{ij})]P_{j1}(I_1(Y_{i1}), \dots, I_{j-1}(Y_{i(j-1)}), I_{j+1}(Y_{i(j+1)}), \dots, I_L(Y_{iL}); F^0; P^{ce0}, P^{ce1})$$
$$+P_{j2}(I_1(Y_{i1}), \dots, I_{j-1}(Y_{i(j-1)}), I_{j+1}(Y_{i(j+1)}), \dots, I_L(Y_{iL}); F^0; P^{ce0}, P^{ce1})\} \cdot \frac{\hat{L}(Y_i)}{g(Y_i)}, \tag{20}$$

where $Y_i = (Y_{i1}, Y_{i2}, \dots, Y_{iL})$. According to (20), the necessary condition for the optimal sensor rules that minimize $C_{MC}(I_1(y_1), \dots, I_L(y_L); F^0; P^{ce0}, P^{ce1}, N)$ is stated in the following lemma:

Lemma 2: Let $\{I_1(y_1), \dots, I_L(y_L)\}$ be the set of optimal sensor rules, i.e., they minimize $C_{MC}(I_1(y_1), \dots, I_L(y_L); F^0; P^{ce0}, P^{ce1}, N)$ in the parallel Bayesian detection network, then they must satisfy the following equations:

$$I_1(Y_{i1}) = I[P_{11}(I_2(Y_{i2}), I_3(Y_{i3}), \dots, I_L(Y_{iL}); F^0; P^{ce0}, P^{ce1}) \cdot \hat{L}(Y_i)], \tag{21}$$

$$I_2(Y_{i2}) = I[P_{21}(I_1(Y_{i1}), I_3(Y_{i3}), \dots, I_L(Y_{iL}); F^0; P^{ce0}, P^{ce1}) \cdot \hat{L}(Y_i)], \tag{22}$$

$$\cdots\cdots$$

$$I_L(Y_{iL}) = I[P_{L1}(I_1(Y_{i1}), I_2(Y_{i2}), \dots, I_{L-1}(Y_{i(L-1)}); F^0; P^{ce0}, P^{ce1}) \cdot \hat{L}(Y_i)], \tag{23}$$

where $P_{j1}(\cdot)$ are defined by (17) and $I[\cdot]$ is an indicator function defined as follows:

$$I[x] = \begin{cases} 1, & \text{if } x \geq 0; \\ 0, & \text{if } x < 0. \end{cases} \tag{24}$$

Proof: Note that both $P_{j1}(\cdot)$ and $P_{j2}(\cdot)$ are independent of $I_j(y_j)$ for $j = 1, \dots, L$. If $I_1(Y_{i1})$ minimizes the Monte Carlo cost function under the given $I_2(Y_{i2}), \dots, I_L(Y_{iL})$, we only need to minimize the first term of the summation in (20), that is, $[1 - I_1(Y_{i1})]P_{11}(I_2(Y_{i2}), I_3(Y_{i3}), \dots, I_L(Y_{iL}); F^0; P^{ce0}, P^{ce1})\frac{\hat{L}(Y_i)}{g(Y_i)}$. Note that the value of $I_1(Y_{i1})$ is 0 or 1 and $g(y)$ is well defined, that is, $g(Y_i) > 0$, $I_1(Y_{i1})$ should be equal to 1 when $P_{11}(I_2(Y_{i2}), I_3(Y_{i3}), \dots, I_L(Y_{iL}); F^0; P^{ce0}, P^{ce1})\hat{L}(Y_i) \geq 0$ for $i = 1, \dots, N$, otherwise it should be equal to 0. Therefore, we obtain (21) by the definition of $I[x]$ in (24). Similarly, we obtain (22) and (23) by minimizing (20).

3.2 An analytical result for the K-out-of-L rule

When the fusion rule is a K-out-of-L rule, we would obtain an analytical result in the presence of nonideal channels. It is described as follows:

Theorem 1.1: If the fusion rule is a K-out-of-L rule and the probabilities of channel errors are less than 0.5 (i.e., $0 < P_j^{ce0} < 0.5$, $0 < P_j^{ce1} < 0.5$) for each channel, the optimal sensor rules are $I_j(Y_{ij}) = I[\hat{L}(Y_i)]$ for $i = 1, \dots, N$ and $j = 1, \dots, L$.

Proof: From Lemma 1, we know

$$P_{j1}(\cdot) = \sum_{k'=1}^{2^L} [1 - F(s_{k'})] P_{m \neq j} \cdot \left(1 - P_j^{ce0} - P_j^{ce1}\right) [1 - 2s_{k'}(j)]$$
$$= \left(1 - P_j^{ce0} - P_j^{ce1}\right) \sum_{k'=1}^{2^{L-1}} [(1 - F(s_{k'}|s_{k'}(j) = 0)) - (1 - F(s_{k'}|s_{k'}(j) = 1))] P_{m \neq j}$$
$$= \left(1 - P_j^{ce0} - P_j^{ce1}\right) \sum_{k'=1}^{2^{L-1}} [F(s_{k'}|s_{k'}(j) = 1) - F(s_{k'}|s_{k'}(j) = 0)] P_{m \neq j}. \tag{25}$$

Since $0 < P_j^{ce0} < 0.5$, $0 < P_j^{ce1} < 0.5$, we have $1 - P_j^{ce0} - P_j^{ce1} > 0$. Obviously, $P_{m \neq j} > 0$ holds from its definition. If $[F(s_{k'}|s_{k'}(j) = 1) - F(s_{k'}|s_{k'}(j) = 0)] \geq 0$, $P_{j1}(\cdot) \geq 0$ can be derived. When the fusion rule is a K-out-of-L rule, $F(s_{k'}) = I\left[\sum_{j=1}^{L} s_{k'}(j) - K\right]$. Thus,

$$F(s_{k'}|s_{k'}(j) = 1) = I\left[\sum_{m=1, m \neq j}^{L} s_{k'}(m) + 1 - K\right],$$
$$F(s_{k'}|s_{k'}(j) = 0) = I\left[\sum_{m=1, m \neq j}^{L} s_{k'}(m) + 0 - K\right].$$

If $\sum_{m=1,m \neq j}^{L} s_{k'}(m) + 0 - K \geq 0$, then $\sum_{m=1,m \neq j}^{L} s_{k'}(m) + 1 - K \geq 0$, and we can get that $F(s_{k'}|s_{k'}(j) = 1) - F(s_{k'}|s_{k'}(j) = 0) = 0$. If $\sum_{m=1,m \neq j}^{L} s_{k'}(m) + 0 - K < 0$, then $F(s_{k'}|s_{k'}(j) = 1) - F(s_{k'}|s_{k'}(j) = 0) \geq 0$. In a word, $F(s_{k'}|s_{k'}(j) = 1) - F(s_{k'}|s_{k'}(j) = 0) \geq 0$ is derived, thus $P_{j1} \geq 0$. It is easy to find a $s_{k'}(m)$, $m \neq j$ so that $\sum_{m=1,m \neq j}^{L} s_{k'}(m) + 0 = K - 1$ and $\sum_{m=1,m \neq j}^{L} s_{k'}(m) + 1 = K$. Thus, there must exist $F(s_{k'}|s_{k'}(j) = 1) - F(s_{k'}|s_{k'}(j) = 0) > 0$. Therefore, the $P_{j1} > 0$ is derived. Recalling the necessary condition for the optimal sensor rules, that is, $I(Y_{ij}) = I[P_{j1}(\cdot) \cdot \hat{L}(Y_i)]$, the proof is completed.

Remark 3: The K-out-of-L rule counts the number of sensors that vote in favor of H_1 and compares it with a given threshold K [37]. It is also referred to as the counting rule or voting rule and is widely used in the practical decision fusion area [38, 39]. It encompasses a general class of fusion rules such as AND, OR, and Majority Boolean fusion rules [40]. The reason we assume that the probabilities of channel errors are less than 0.5 is based on practical considerations. If the probabilities of channel errors are greater than or equal to 0.5, the channel is totally unreliable and the performance is not better than a random decision. Obviously, the analytical solution is very efficient to tackle large-scale sensor networks with dependent observations and channel errors.

4. Monte Carlo Gauss-Seidel iterative algorithm and its convergence

For general fusion rules that do not have the form of a K-out-of-L rule, an efficient algorithm can be obtained that is inspired by Lemma 2. Next, we present a Monte Carlo Gauss-Seidel iterative algorithm and derive its convergence, when the fusion rule is fixed.

4.1 Monte Carlo Gauss-Seidel iterative algorithm

Based on the fixed-point type necessary condition given in Lemma 2, the Monte Carlo Gauss-Seidel iterative algorithm is presented in Algorithm 1.

Algorithm 1: Optimization of the sensor rules.

Given the fusion rule F^0:

- Step 1: Generate N samples: $Y_1, \ldots, Y_N \sim g(y)$, where $g(y)$ is an importance sampling density and $Y_i = [Y_{i1}, Y_{i2}, \ldots, Y_{iL}]$.
- Step 2: Initialize the L sensor rules, for $j = 1, 2, \ldots, L$ and $i = 1, \ldots, N$,

$$I_j^{(0)}(Y_{ij}) = 0/1. \tag{26}$$

- Step 3: Iteratively search for the L sensor rules until a termination criterion in Step 4 is satisfied. The $n+1$th iteration is given as follows: for $i = 1, \ldots, N$,

$$I_1^{(n+1)}(Y_{i1}) = I\left[P_{11}\left(I_2^{(n)}(Y_{i2}), I_3^{(n)}(Y_{i3}), \ldots, I_L^{(n)}(Y_{iL}); F^0; P^{ce0}, P^{ce1}\right)\hat{L}(Y_i\right], \tag{27}$$

$$I_2^{(n+1)}(Y_{i2}) = I\left[P_{21}\left(I_1^{(n+1)}(Y_{i1}), I_3^{(n)}(Y_{i3}), \ldots, I_L^{(n)}(Y_{iL}); F^0; P^{ce0}, P^{ce1}\right)\hat{L}(Y_i)\right], \tag{28}$$

$$I_L^{(n+1)}(Y_{iL}) = I\left[P_{L1}\left(I_1^{(n+1)}(Y_{i1}), \ldots, I_{L-1}^{(n+1)}(Y_{i(L-1)}); F^0; P^{ce0}, P^{ce1}\right)\hat{L}(Y_i)\right]. \tag{29}$$

- Step 4: For $i = 1, \ldots, N$, the termination criterion of the iteration process is

$$\begin{aligned} I_1^{(n+1)}(Y_{i1}) &= I_1^{(n)}(Y_{i1}), \\ I_2^{(n+1)}(Y_{i2}) &= I_2^{(n)}(Y_{i2}), \\ &\cdots\cdots \\ I_L^{(n+1)}(Y_{iL}) &= I_L^{(n)}(Y_{iL}). \end{aligned} \tag{30}$$

Remark 4: When we obtain $I_1(Y_{i1})$ for $i = 1, \ldots, N$, we can compress y_1 by defining $I_1(y_1) = I_1(Y_{i1})$ if the distance $\|y_1 - Y_{i1}\| \le \|y_1 - Y_{i'1}\|$ for all $i' \neq i$. In the same way, we can compress y_j for $j = 2, \ldots, L$. In fact, the method is to find one nearest neighbor of y_j for $j = 1, \ldots, L$ and use the corresponding compression rule. Moreover, we can utilize the k-nearest neighbor (**knn**) to compress y_j (see more in [41]).

Remark 5: The main computation burden of Algorithm 1 is included in (27)–(29). If we let the number of discretized points $N_1 = N_2 = \ldots = N_L = N$ in [31], then $P_{j1}(\cdot)\hat{L}(Y_i), j = 1, \ldots, L$, and $i = 1, \ldots, N$ are computed LN^L times for each iteration, as in [31]. But they only need to be computed LN times in Algorithm 1. Thus, the computational complexity of Algorithm 1, i.e., $O(LN)$ is much less than that in [31], that is, $O(LN^L)$. It is more efficient to tackle large-scale sensor networks with dependent observations and channel errors.

4.2 Convergence of the iterative algorithm

Now, we show that Algorithm 1 must converge to a stationary point and the algorithm cannot oscillate infinitely, that is, it terminates after a finite number of iterations.

Lemma 3: Given the fusion rule F^0, for any initial values $\left(I_1^{(0)}, \dots, I_L^{(0)}\right)$ in (26), $C_{MC}\left(I_1^{(n)}, \dots, I_j^{(n)}, I_{j+1}^{(n)}, \dots, I_L^{(n)}; F^0; P^{ce0}, P^{ce1}, N\right)$ must converge to a stationary point after a finite number of iterations.

Proof: For $j = 1, \dots, L$, we denote C_{MC} (20) in the $n + 1$th iteration process by

$$C_{MC}\left(I_1^{(n+1)}, \dots, I_j^{(n+1)}, I_{j+1}^{(n)}, \dots, I_L^{(n)}; F^0; P^{ce0}, P^{ce1}, N\right)$$
$$= \frac{1}{N}\sum_{i=1}^{N}\Big\{\left[1 - I_j^{(n+1)}(Y_{ij})\right]P_{j1}(I_1^{(n+1)}(Y_{i1}), \dots, I_{j-1}^{(n+1)}(Y_{i(j-1)}), I_{j+1}^{(n)}(Y_{i(j+1)}), \dots, I_L^{(n)}(Y_{iL}); F^0;$$
$$P^{ce0}, P^{ce1}, N) + P_{j2}(I_1^{(n+1)}(Y_{i1}), \dots, I_{j-1}^{(n+1)}(Y_{i(j-1)}), I_{j+1}^{(n)}(Y_{i(j+1)}), \dots, I_L^{(n)}(Y_{iL}); F^0; P^{ce0}, P^{ce1}, N)\Big\}$$
$$\cdot\frac{\hat{L}(Y_i)}{g(Y_i)} + c. \tag{31}$$

Similarly, we denote the $(n + 1)$th iteration process of the iterative items $P_{j1}(\cdot)\hat{L}(\cdot)$ in (27)–(29) by

$$G_j^i = P_{j1}\left(I_1^{(n+1)}(Y_{i1}), \dots, I_{j-1}^{(n+1)}(Y_{i(j-1)}), I_{j+1}^{(n)}(Y_{i(j+1)}), \dots, I_L^{(n)}(Y_{iL}); F^0; P^{ce0}, P^{ce1}, N\right)\hat{L}(Y_i), \tag{32}$$

for $i = 1, \dots, N$ and $j = 1, \dots, L$. Plugging G_j^i into (31), we know

$$C_{MC}\left(I_1^{(n+1)}, \dots, I_j^{(n+1)}, I_{j+1}^{(n)}, \dots, I_L^{(n)}; F^0; P^{ce0}, P^{ce1}, N\right) = \frac{1}{N}\sum_{i=1}^{N}\frac{\left[1 - I_j^{(n+1)}(Y_{ij})\right]}{g(Y_i)}G_j^i + C_j^i, \tag{33}$$

where $C_j^i = c + \frac{1}{N}\sum_{i=1}^{N}P_{j2}\left(I_1^{(n+1)}(Y_{i1}), \dots, I_{j-1}^{(n+1)}(Y_{i(j-1)}), I_{j+1}^{(n)}(Y_{i(j+1)}), \dots, I_L^{(n)}(Y_{iL}); F^0; P^{ce0}, P^{ce1}, N\right)\frac{\hat{L}(Y_i)}{g(Y_i)}$ is independent of $I_j^{(n)}$ and $I_j^{(n+1)}$. Splitting $1 - I_j^{(n+1)}(Y_{ij})$ into two terms, we obtain

$$\begin{aligned}
&C_{MC}\left(I_1^{(n+1)}, \dots, I_j^{(n+1)}, I_{j+1}^{(n)}, \dots, I_L^{(n)}; F^0; P^{ce0}, P^{ce1}, N\right)\\
&= \frac{1}{N}\sum_{i=1}^{N}\frac{\left[1 - I_j^{(n)}(Y_{ij})\right] + \left[I_j^{(n)}(Y_{ij}) - I_j^{(n+1)}(Y_{ij})\right]}{g(Y_i)}G_j^i + C_j^i\\
&= \frac{1}{N}\sum_{i=1}^{N}\frac{\left[1 - I_j^{(n)}(Y_{ij})\right]}{g(Y_i)}G_j^i + C_j^i + \frac{1}{N}\sum_{i=1}^{N}\frac{\left[I_j^{(n)}(Y_{ij}) - I_j^{(n+1)}(Y_{ij})\right]}{g(Y_i)}G_j^i\\
&= C_{MC}\left(I_1^{(n+1)}, \dots, I_{j-1}^{(n+1)}, I_j^{(n)}, \dots, I_L^{(n)}; F^0; P^{ce0}, P^{ce1}, N\right) + D_j^{(n+1)},
\end{aligned} \tag{34}$$

where

$$D_j^{(n+1)} = \frac{1}{N}\sum_{i=1}^{N} \frac{\left[I_j^{(n)}(Y_{ij}) - I_j^{(n+1)}(Y_{ij})\right]}{g(Y_i)} G_j^i. \tag{35}$$

Note that (27)–(29) imply that $I_j^{(n+1)}(Y_{ij}) = 0$ if and only if $G_j^i < 0$ and $I_j^{(n+1)}(Y_{ij}) = 1$ if and only if $G_j^i \geq 0$ for $i = 1, \ldots, N, j = 1, \ldots, L$. It means

$$\left[I_j^{(n)}(Y_{ij}) - I_j^{(n+1)}(Y_{ij})\right] G_j^i \leq 0. \tag{36}$$

Thus, for $\forall i, j$

$$\left[I_j^{(n)}(Y_{ij}) - I_j^{(n+1)}(Y_{ij})\right] G_j^i / g(Y_i) \leq 0, \tag{37}$$

where the inequality (35) holds since $g(\cdot)$ is well-defined (i.e., $g(\cdot) > 0$). Substituting (35) into (33) yields $D_j^{(n+1)} \leq 0$. Thus, for $\forall j \leq L$,

$$\begin{aligned} &C_{MC}\left(I_1^{(n+1)}, \ldots, I_j^{(n+1)}, I_{j+1}^{(n)}, \ldots, I_L^{(n)}; F^0; p^{ce0}, p^{ce1}, N\right) \\ &\leq C_{MC}\left(I_1^{(n+1)}, \ldots, I_{j-1}^{(n+1)}, I_j^{(n)}, \ldots, I_L^{(n)}; F^0; p^{ce0}, p^{ce1}, N\right). \end{aligned} \tag{38}$$

Furthermore,

$$\begin{aligned} &C_{MC}\left(I_1^{(n+1)}, I_2^{(n+1)}, \ldots, I_L^{(n+1)}; F^0; p^{ce0}, p^{ce1}, N\right) \\ &\leq C_{MC}\left(I_1^{(n)}, I_2^{(n)}, \ldots, I_L^{(n)}; F^0; p^{ce0}, p^{ce1}, N\right). \end{aligned} \tag{39}$$

It means C_{MC} is nonincreasing. Note that $C_{MC}\left(I_1^{(n)}, I_2^{(n)}, \ldots, I_L^{(n)}; F^0; p^{ce0}, p^{ce1}, N\right)$ is a finite value. We conclude that it must converge to a stationary point after a finite number of iterations.

Theorem 1.2: Given the fusion rule F^0, the sensor rules $I_1^{(n)}, I_2^{(n)}, \ldots, I_L^{(n)}$ are finitely convergent, i.e., Algorithm 1 converges after a finite number of iterations.

Proof: By Lemma 3, C_{MC} must attain a stationary point after a finite number of iterations. It means that the value of C_{MC} cannot change after nth iteration, that is,

$$\begin{aligned} &C_{MC}\left(I_1^{(n+1)}, \ldots, I_j^{(n+1)}, I_{j+1}^{(n)}, \ldots, I_L^{(n)}; F^0; p^{ce0}, p^{ce1}, N\right) \\ &= C_{MC}\left(I_1^{(n+1)}, \ldots, I_{j-1}^{(n+1)}, I_j^{(n)}, \ldots, I_L^{(n)}; F^0; p^{ce0}, p^{ce1}, N\right). \end{aligned} \tag{40}$$

Using (32) and (37), we derive that $D_j^{(n+1)} = 0$. Combining (33)–(35), we know

$$\left[I_j^{(n)}(Y_{ij}) - I_j^{(n+1)}(Y_{ij})\right] G_j^i = 0, \textit{for } i = 1, \ldots, N, \tag{41}$$

which implies either $I_j^{(n)}(Y_{ij}) - I_j^{(n+1)}(Y_{ij}) = 0$, *i.e.*, $I_j^{(n)}(Y_{ij}) = I_j^{(n+1)}(Y_{ij})$ or $G_j^i = 0$, *i.e.*, $I_j^{(n+1)}(Y_{ij}) = 1, I_j^{(n)}(Y_{ij}) = 0$. It follows that when C_{MC} converges to a stationary point, either $I_j^{(n+1)}(Y_{ij})$ is invariant or $I_j^{(n+1)}(Y_{ij}) = 1, I_j^{(n)}(Y_{ij}) = 0$. Namely, $I_j^{(n+1)}(Y_{ij})$ can only change from 0 to 1 at most a finite number of times. Therefore, the $I_1^{(n)}, I_2^{(n)}, \ldots, I_L^{(n)}$ are finitely convergent.

5. Extension for simultaneous search for the optimal sensor rules and fusion rule

In this section, we extend the Monte Carlo method to search for the optimal sensor rules and the optimal fusion rule simultaneously. Firstly, the necessary condition is generalized to search for the optimal sensor rules and the optimal fusion rule simultaneously. Secondly, we describe a generalized Monte Carlo Gauss-Seidel iterative algorithm. We also give the convergence of the iterative algorithm.

5.1 A necessary condition for the optimal sensor rules and the optimal fusion rule

Note that (15) can be rewritten as follows:

$$\begin{aligned} &C_{MC}(I_1(y_1), \ldots, I_L(y_L); F^0, ; P^{ce0}, , P^{ce1}, N) \\ &= c + \frac{1}{N}\sum_{i=1}^{N}\sum_{k'=1}^{2^L}\sum_{k=1}^{2^L}\left[1 - F^0(s_{k'})\right]P(s_{k'}|s_k) \cdot P_{s_k}(\mathbf{I}(Y_i))\frac{\hat{L}(Y_i)}{g(Y_i)} \\ &= c + \frac{1}{N}\sum_{k'=1}^{2^L}\left[1 - F^0(s_{k'})\right]\sum_{i=1}^{N}\sum_{k=1}^{2^L}P(s_{k'}|s_k) \cdot P_{s_k}(\mathbf{I}(Y_i))\frac{\hat{L}(Y_i)}{g(Y_i)}, \end{aligned} \tag{42}$$

where $P_{s_k}(\mathbf{I}(Y_i)) \triangleq \prod_{j=1}^{L}\left[s_k(j)I_j(Y_{ij}) + (1 - s_k(j))(1 - I_j(Y_{ij}))\right]$ and $\mathbf{I}(Y_i) = (I_1(Y_{i1}), I_2(Y_{i2}), \ldots, I_L(Y_{iL}))$. Since $P_{s_k}(\mathbf{I}(Y_i) = 1$ if and only if $I_j = s_k(j)$ for all $j = 1, \ldots, L$, (39) can be simplified as follows:

$$\begin{aligned} &C_{MC}(I_1(y_1), \ldots, I_L(y_L); F^0, ; P^{ce0}, , P^{ce1}, N) \\ &= c + \frac{1}{N}\sum_{k'=1}^{2^L}\left[1 - F^0(s_{k'})\right] \cdot \sum_{i=1}^{N}P(s_{k'}|(I_1(Y_{i1}), \ldots, I_L(Y_{iL})))\frac{\hat{L}(Y_i)}{g(Y_i)}, \end{aligned} \tag{43}$$

where the terms $P_{s_k}(\mathbf{I}(Y_i)) = 0$ are eliminated.

Remark 6: According to (20) and (40), the necessary condition for the optimal sensor rules is similar to Lemma 2 and the necessary condition for the optimal fusion rule is given by

$$F^0(s_{k'}) = I\left[\sum_{i=1}^{N}P(s_{k'}|(I_1(Y_{i1}), \ldots, I_L(Y_{iL}))) \cdot \frac{\hat{L}(Y_i)}{g(Y_i)}\right] \tag{44}$$

for $k' = 1, \ldots, 2^L$. The proofs are similar to Lemma 2.

5.2 Generalized Monte Carlo Gauss-Seidel iterative algorithm

Based on the fixed-point type necessary condition, the generalized Monte Carlo Gauss-Seidel iterative algorithm is presented in Algorithm 2.

Remark 7: For any initial values $\left(I_1^{(0)}, \dots, I_L^{(0)}; F^{0(0)}\right)$, the Monte Carlo cost function $C_{MC}\left(I_1^{(n)}, \dots, I_j^{(n)}, I_{j+1}^{(n)}, \dots, I_L^{(n)}; F^{0(n)}; P^{ce0}, P^{ce1}, N\right)$ must converge to a stationary point and Algorithm 2 terminates after a finite number of iterations. The proofs are similar to those of Lemma 3 and Theorem 1.2.

Algorithm 2: Simultaneous optimization of the sensor rules and the fusion rule.

- Step 1: Generate N samples: $Y_1, \dots, Y_N \sim g(y)$, where $g(y)$ is an importance sampling density and $Y_i = [Y_{i1}, Y_{i2}, \dots, Y_{iL}]$.

- Step 2: Initialize the L sensor rules and the fusion rule, respectively, for $j = 1, 2, \dots, L$, $i = 1, \dots, N$, and $k' = 1, \dots, 2^L$,

$$I_j^{(0)}(Y_{ij}) = 0/1,\ F^{0(0)}(s_{k'}) = 0/1.$$

- Step 3: Iteratively search for the L sensor rules and the fusion rule until a termination criterion in Step 4 is satisfied. The $n+1$th iteration is given as follows: for $i = 1, \dots, N$ and $k' = 1, \dots, 2^L$

$$I_1^{(n+1)}(Y_{i1}) = I\left[P_{11}\left(I_2^{(n)}(Y_{i2}), I_3^{(n)}(Y_{i3}), \dots, I_L^{(n)}(Y_{iL}); F^{0(n)}; P^{ce0}, P^{ce1}\right) \cdot \hat{L}(Y_i)\right],$$

$$I_2^{(n+1)}(Y_{i2}) = I\left[P_{21}\left(I_1^{(n+1)}(Y_{i1}), I_3^{(n)}(Y_{i3}), \dots, I_L^{(n)}(Y_{iL}); F^{0(n)}; P^{ce0}, P^{ce1}\right) \cdot \hat{L}(Y_i)\right],$$

$$\dots\dots$$

$$I_L^{(n+1)}(Y_{iL}) = I\left[P_{L1}\left(I_1^{(n+1)}(Y_{i1}), I_2^{(n+1)}(Y_{i2}), \dots, I_{L-1}^{(n+1)}(Y_{i(L-1)}); F^{0(n)}; P^{ce0}, P^{ce1}\right) \cdot \hat{L}(Y_i)\right],$$

$$F^{0(n+1)}(s_{k'}) = I\left[\sum_{i=1}^{N} P\left(s_{k'} \middle| \left(I_1^{(n+1)}(Y_{i1}), I_2^{(n+1)}(Y_{i2}), \dots, I_L^{(n+1)}(Y_{iL})\right)\right)\frac{\hat{L}(Y_i)}{g(Y_i)}\right].$$

- Step 4: For $i = 1, \dots, N$ and $k' = 1, 2, \dots, 2^L$, the termination criterion of the iteration process is

$$I_1^{(n+1)}(Y_{i1}) = I_1^{(n)}(Y_{i1}),$$

$$I_2^{(n+1)}(Y_{i2}) = I_2^{(n)}(Y_{i2}),$$

$$\dots\dots$$

$$I_L^{(n+1)}(Y_{iL}) = I_L^{(n)}(Y_{iL});$$

$$F^{0(n+1)}(s_{k'}) = F^{0(n)}(s_{k'}).$$

6. Numerical examples

In this section, in order to evaluate the performance of Algorithms 1 and 2, we present some examples with a Gaussian signal s observed in the presence of Gaussian sensor noises.

The random signal s and observation noises $v_1, v_2, \dots, v_L$ are as follows:

$$H_0: \ y_j = v_j; \qquad H_1: \ y_j = s + v_j, \qquad for\, j = 1, \dots, L, \tag{45}$$

where $v_1, v_2 \dots, v_L, s$ are all mutually independent and

$$v_j \sim N(0, 0.6), \quad s \sim N(1, 0.4), \quad for\, j = 1, \dots, L.$$

Thus, given H_0 and H_1, the two conditional probability density functions are

$$p(y_1, y_2, \dots, y_L|H_0) \sim N(\mu_0, \Sigma_0), \qquad p(y_1, y_2, \dots, y_L|H_1) \sim N(\mu_1, \Sigma_1),$$

where $\mu_0, \mu_1, \Sigma_0, \Sigma_1$ are easily obtained from the relationship of $s, v_1, v_2, \dots, v_L$.

Assume that each sensor is required to transmit a bit through a channel with probabilities of $P_j^{ceo} = P_j^{ce1} = p$, where $p = 0.05, 0.15, 0.3$, for $j = 1, 2, \dots, L$. In the cost function (2), let the cost coefficients $C_{00} = C_{11} = 0$ and $C_{10} = C_{01} = 1$. The receiver operating characteristics (ROC) curves are used to evaluate the performance of the algorithms. Pf and Pd denote the probability of false alarm and the probability of detection, respectively.

6.1 Two-sensor network

We compare the Monte Carlo Gauss-Seidel iterative algorithm with the centralized algorithm and the iterative algorithm based on the Riemann sum approximation in [31] by using the receiver operating characteristics (ROC) curves.

In this case, we know $\mu_0 = [0, 0]^T$, $\mu_1 = [1, 1])^T$ and $\Sigma_0 = [0.6, 0; 0, 0.6]$, $\Sigma_1 = [1, 0.4; 0.4, 1]$. Some discrete values of a and b are used to plot ROC curves. We refer to the optimal importance sampling density $g(y) \propto |P_{H_0^0}(y)\hat{L}(y)|$ in Section 2.2 and $|\hat{L}(y)| = |ap(y|H_1) - bp(y|H_0)|$. The form is similar to the mixture-Gaussian distribution. Therefore, the importance sampling density $g(y)$ is chosen to be the mixture-Gaussian distribution. The effects of choosing different $g(y)$ in terms of the performance of the Monte Carlo method were shown in [21] *via* numerical examples. For Algorithm 1, we take $N = 200$ samples from the density $g(y)$. For the Riemann sum approximation iterative algorithm in [31], we take a discretized step-size $\Delta = 0.09$, $y_i \in [-8, 10]$, i.e., $N_1 = N_2 = N = 200$. The ROC curves for three important fusion rules: AND, OR, and XOR rules with $p = 0.05, 0.15, 0.3$ are plotted in **Figure 2**. We compare the computational time of the two algorithms with $p = 0.15$ in **Figure 3**. Note that the analytical solution is used for the AND rule and the OR rule. Since the XOR rule is not a K-out-of-L rule, we use Algorithm 1 to search for the sensor rules.

Some observations in **Figures 2** and **3** are presented as follows:

- Given the fusion rule, the two points $(0, 0)$ and $(1, 1)$ may not be the beginning or ending points of the ROC curves, which is different from the case in the ideal channel cases. In addition, the larger the probability of channel errors is, the

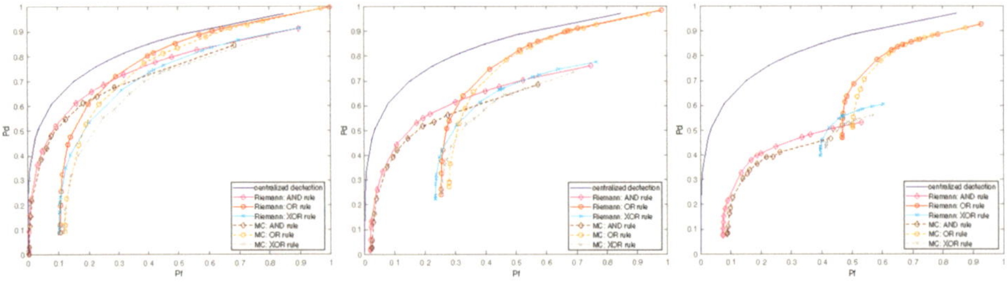

Figure 2.
Two-sensor ROC curves with the probability of channel errors $p = 0.05, 0.15, 0.3$.

farther away from (0, 0) or (1, 1) the ROC curves are. A possible reason is that the detection probability is not equal to 0 or 1, even when the false alarm probability is 0 or 1 in the presence of channel errors.

- From **Figure 2**, when the probability of channel transmission errors increases, the decision fusion performance of different methods using the optimal sensor rules decreases.

- It can be seen in **Figure 2** that the ROC curves of the new Monte Carlo approach are very close to those of the previous algorithm based on the Riemann sum approximation. However, from **Figure 3**, the computational time of the Monte Carlo importance sampling approximation is much less than that of the Riemann sum approximation for the three different fusion rules. It also implies that the new method can be used to deal with large-scale sensor networks.

- Note that the computational time of the AND rule and the OR rule is less than that of the XOR rule for the Monte Carlo importance sampling approximation from **Figure 3**. The reason is that the AND rule and the OR rule belong to the K-out-of-L rules. The analytical form is used for the AND rule and the OR rule, therefore, the corresponding computation time is relatively lower.

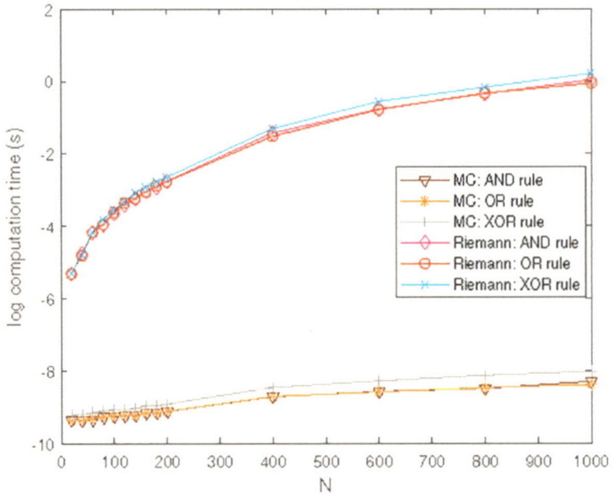

Figure 3.
Two-sensor computational time as N increases with the probability of channel errors $p = 0.15$.

6.2 Ten-sensor network

We consider a larger sensor network with 10 sensors, which cannot be dealt with by the previous decision fusion algorithm based on the Riemann sum approximation due to its heavy computation requirements. For different probabilities of channel errors, the ROC curves of the AND rule, the OR rule, the 4-out-of-10 rule, the 6-out-of-10 rule, and Algorithm 2 are plotted in **Figure 4**.

Some observations in **Figure 4** are presented as follows:

- The ROC curves for the ten-sensor network exhibit similar behavior as those for the two-sensor network.
- Given the fusion rule and the probability of channel errors, the decision fusion performance of the AND rule is better than the other rules for a small false alarm probability and the decision fusion performance of the OR rule is better than the other rules for a large false alarm probability. The reason may be that both of them are extreme cases of the fusion rules. For other cases, the 4-out-of-10 rule and the 6-out-of-10 rule perform better than the AND rule and the OR rule, respectively.
- Regardless of the centralized detection algorithm, **Figure 4** shows that the ROC curves generated by Algorithm 2 obtain almost the best performance for different probabilities of channel errors. It implies that a simultaneous search for the sensor rules and the fusion rule would provide better decision fusion performance.

6.3 One-hundred-sensor network

We consider a large-scale network with one hundred sensors. The parameter settings are similar to Section 6.2. The ROC curves of the 20-out-of-100 rule, the 40-out-of-100, the 50-out-of-100, the 60-out-of-100 rule, and the 80-out-of-100 rule are plotted in **Figure 5**.

From **Figure 5**, it can be seen that the dramatically lower computational requirement of our method enables us to handle a large sensor network consisting of one hundred sensors. This is due to the fact that we have shown that there exist analytical forms of the optimal sensor rules for the K-out-of-L rule. In addition, the decision fusion performance of different methods is improved, as the number of sensors becomes large.

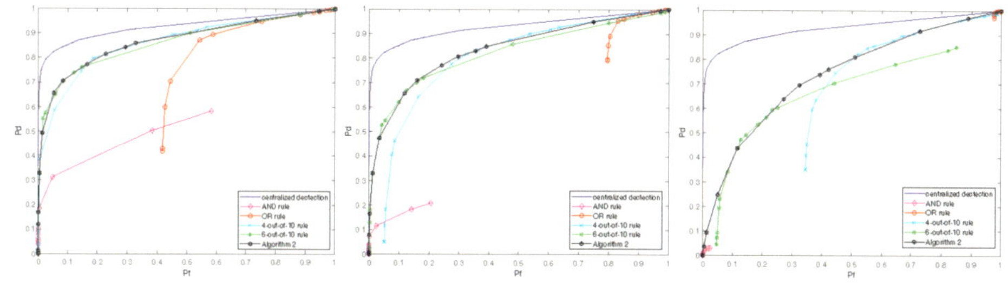

Figure 4.
Ten-sensor ROC curves with the probability of channel errors $p = 0.05, 0.15, 0.3$.

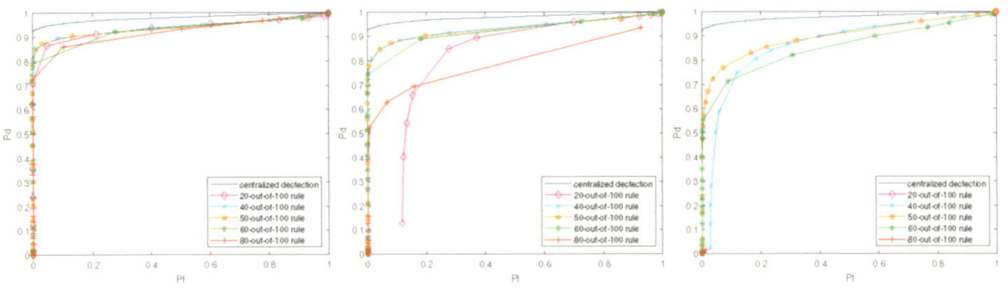

Figure 5.
One-hundred-sensor ROC curves with the probability of channel errors $p = 0.05, 0.15, 0.3$.

7. Conclusion

By employing the Monte Carlo importance sampling technique, decision fusion algorithms have been provided for large-scale sensor networks with dependent observations and channel errors. The Bayesian cost function is approximated by the Monte Carlo cost function. The necessary conditions for the optimal sensor rules and the optimal fusion rule that minimize the Monte Carlo cost function have been obtained. Computationally efficient Monte Carlo Gauss-Seidel iterative algorithms have been proposed to search for the optimal sensor rules and the optimal fusion rule. These algorithms have been shown to converge after a finite number of iterations. The computational complexity of the new algorithm (i.e., $O(LN)$) is much less than that of the previous algorithm based on Riemann sum approximation (i.e., $O(LN^L)$). For the K-out-of-L rule, an analytical solution has been presented for the optimal sensor rules. Simulations have demonstrated the effectiveness of Algorithms 1 and 2. Future work will include the decision fusion algorithms under the Monte Carlo framework for other networks such as tandem networks, tree networks, and sensor networks under Byzantine attack.

Acknowledgements

This work was supported in part by the Sichuan Youth Science and Technology Innovation Team under Grant 2022JDTD0014, and Grant 2021JDJQ0036.

Author details

Yiwei Liao[1,2], Xiaojing Shen[3]*, Junfeng Wang[4] and Yunmin Zhu[3]

1 School of Data Science, The Chinese University of Hong Kong, Shenzhen, Guangdong, China

2 Shcool of Information Science and Technology, University of Science and Technology of China, Hefei, Anhui, China

3 School of Mathematics, Sichuan University, Chengdu, Sichuan, China

4 School of Computer Science, Sichuan University, Chengdu, Sichuan, China

*Address all correspondence to: shenxj@scu.edu.cn

References

[1] Blum RS, Kassam SA, Vincent H, Poor. Distributed detection with multiple sensors: Part II-advanced topics. Proceedings of the IEEE. 1997; **85**(1):64-79

[2] Chair Z, Varshney PK. Optimal data fusion in multiple sensor detection systems. IEEE Transactions on Aerospace and Electronic Systems. 1986; **22**(1):98-101

[3] Li C, Li G, Varshney PK. Distributed detection of sparse stochastic signals with 1-bit data in tree-structured sensor networks. IEEE Transactions on Signal Processing. 2020;**68**:2963-2976

[4] Robert R Tenney and Nils R Sandell. Detection with distributed sensors. IEEE Transactions on Aerospace and Electronic Systems. 1981;**17**(4):501–510

[5] Tsitsiklis J, Athans M. On the complexity of decentralized decision making and detection problems. IEEE Transactions on Automatic Control. 1985;**30**(5):440-446

[6] Varshney PK. Distributed Detection and Data Fusion. New York: Springer-Verlag; 1997

[7] Viswanathan R, Varshney PK. Distributed detection with multiple sensors: Part I-fundamentals. Proceedings of the IEEE. 1997;**85**(1):54-63

[8] Chen H, Chen B, Varshney PK. Further results on the optimality of the likelihood-ratio test for local sensor decision rules in the presence of nonideal channels. IEEE Transactions on Information Theory. 2009;**55**(2): 828-832

[9] Hoballah IY, Varshney PK. Distributed Bayesian signal detection. IEEE Transactions on Information Theory. 1989;**35**(5):995-1000

[10] Reibman AR, Nolte LW. Optimal detection and performance of distributed sensor systems. IEEE Transactions on Aerospace and Electronic Systems. 2010;**23**(1):24-30

[11] Blum RS, Kassam SA. Optimum distributed detection of weak signals in dependent sensors. IEEE Transactions on Information Theory. 1992;**38**(3): 1066-1079

[12] Tang Z-B, Pattipati KR, Kleinman DL. An algorithm for determining the decision thresholds in a distributed detection problem. IEEE Transactions on Systems, Man, and Cybernetics. 1991;**21**(1):231-237

[13] Tang Z-B, Pattipati KR, Kleinman DL. A distributed m-ary hypothesis testing problem with correlated observations. IEEE Transactions on Automatic Control. 1992;**37**(7):1042-1046

[14] Willett PK, Swaszek PF, Blum RS. The good, bad and ugly: Distributed detection of a known signal in dependent Gaussian noise. IEEE Transactions on Signal Processing. 2000;**48**(12): 3266-3279

[15] Zhu Y, Blum RS, Luo Z-Q, Wong KM. Unexpected properties and optimum-distributed sensor detectors for dependent observation cases. IEEE Transactions on Automatic Control. 2000;**45**(1):62-72

[16] Chen H, Chen B, Varshney PK. A new framework for distributed detection with conditionally dependent observations. IEEE Transactions on Signal Processing. 2012;**60**(3):1409-1419

[17] Ozdemir O, Allen T, Choi S, Wimalajeewa T, Varshney PK. Copula based classifier fusion under statistical dependence. In: IEEE Transactions on Pattern Analysis and Machine Intelligence. 2017

[18] Sundaresan A, Varshney PK, Rao NSV. Copula-based fusion of correlated decisions. IEEE Transactions on Aerospace and Electronic Systems. 2011;**47**(1):454-471

[19] Zhang S, Khanduri P, Varshney PK. Distributed sequential detection: Dependent observations and imperfect communication. IEEE Transactions on Signal Processing. 2020;**68**:830-842

[20] Liao Y, Shen X, Rao H. Analytic sensor rules for optimal distributed decision given *K*-out-of-*L* fusion rule under Monte Carlo approximation. IEEE Transactions on Automatic Control. 2020;**65**(12):5488-5495

[21] Rao H, Shen X, Zhu Y, Pan J. Distributed detection fusion via Monte Carlo importance sampling. In: 35th Chinese Control Conference. Chengdu, China: IEEE; 2016. pp. 4830-4835

[22] Drakopoulos ELIAS, Lee C-C. Optimum multisensor fusion of correlated local decisions. IEEE Transactions on Aerospace and Electronic Systems. 1991;**27**(4):593-606

[23] Kam M, Zhu Q, Steven W, Gray. Optimal data fusion of correlated local decisions in multiple sensor detection systems. IEEE Transactions on Aerospace and Electronic Systems. 1992; **28**(3):916-920

[24] Quan Z, Ma WK, Cui S, Sayed AH. Optimal linear fusion for distributed detection via semidefinite programming. IEEE Transactions on Signal Processing. 2010;**58**(4):2431-2436

[25] Zhu Y, Li X. Unified fusion rules for multisensor multihypothesis network decision systems. IEEE Transactions on Systems, Man, and Cybernetics-Part A: Systems and Humans. 2003;**33**(4): 502-513

[26] Zhu Y, Zhou J, Shen X, Song E, Luo Y. Networked Multisensor Decision and Estimation Fusion: Based on Advanced Mathematical Methods. Boca Raton: CRC Press; 2012

[27] Shen X, Zhu Y, He L, You Z. A near-optimal iterative algorithm via alternately optimizing sensor and fusion rules in distributed decision systems. IEEE Transactions on Aerospace and Electronic Systems. 2011;**47**(4): 2514-2529

[28] Chen B, Jiang R, Kasetkasem T, Varshney PK. Channel aware decision fusion in wireless sensor networks. IEEE Transactions on Signal Processing. 2004; **52**(12):3454-3458

[29] Chen B, Willett PK. On the optimality of the likelihood-ratio test for local sensor decision rules in the presence of nonideal channels. IEEE Transactions on Information Theory. 2005;**51**(2):693-699

[30] Niu R, Chen B, Varshney PK. Fusion of decisions transmitted over Rayleigh fading channels in wireless sensor networks. IEEE Transactions on Signal Processing. 2006;**54**(3):1018-1027

[31] Ren Q, Zhu Y, Shen X, Song E. Optimal sensor rules and unified fusion rules for multisensor multi-hypothesis network decision systems with channel errors. Automatica. 2009;**45**(7): 1694-1702

[32] Thomopoulos SCA, Zhang L. Distributed decision fusion in the presence of networking delays and

channel errors. Information Sciences. 1992;**66**(1–2):91-118

[33] Yilmaz Y, Moustakides GV, Wang X. Channel-aware decentralized detection via level-triggered sampling. IEEE Transactions on Signal Processing. 2012; **61**(2):300-315

[34] Liu JS. Monte Carlo Strategies in Scientific Computing. New York: Springer Science & Business Media; 2008

[35] Christian P. Robert and George Casella. Monte Carlo Statistical Methods. 2nd ed. New York: Springer; 2004

[36] Liao Y, Shen X, Zhu Y. Distributed detection fusion with nonideal channels under Monte Carlo framework. In: 20th International Conference on Information Fusion (Fusion). Xi'an, China: IEEE; 2017. pp. 1-8

[37] Vergara L. On the equivalence between likelihood ratio tests and counting rules in distributed detection with correlated sensors. Signal Processing. 2007;**87**(7):1808-1815

[38] Abdelhakim M, Lightfoot LE, Ren J, Li T. Distributed detection in mobile access wireless sensor networks under byzantine attacks. IEEE Transactions on Parallel and Distributed Systems. 2014; **25**(4):950-959

[39] Viswanathan R, Aalo V. On counting rules in distributed detection. IEEE Transactions on Acoustics, Speech, and Signal Processing. 1989;**37**(5): 772-775

[40] Chaudhari S, Lundén J, Koivunen V, Vincent H, Poor. Bep walls for cooperative sensing in cognitive radios using K-out-of-N fusion rules. Signal Processing. 2013;**93**(7): 1900-1908

[41] Friedman J, Hastie T, Tibshirani R. The Elements of Statistical Learning: Data Mining, Inference, and Prediction. 2nd ed. Boca Raton: Springer; 2009